15th Anglo-French Physical Acoustics Conference (AFPAC 2016)

Journal of Physics: Conference Series Volume 797

Surrey, United Kingdom
13 - 15 January 2016

ISBN: 978-1-5108-3561-0
ISSN: 1742-6588

Printed from e-media with permission by:

Curran Associates, Inc.
57 Morehouse Lane
Red Hook, NY 12571

proceedings.com

Some format issues inherent in the e-media version may also appear in this print version.

Printed by Curran Associates, Inc. (2017)

Institute of Physics
Dirac House, Temple Back
Bristol BS1 6BE UK

Phone: 44 1 17 929 7481
Fax: 44 1 17 920 0979

techtracking@iop.org

Additional copies of this publication are available from:

Curran Associates, Inc.
57 Morehouse Lane
Red Hook, NY 12571 USA
Phone: 845-758-0400
Fax: 845-758-2633
Email: curran@proceedings.com
Web: www.proceedings.com

Table of contents

Volume 797

15th Anglo-French Physical Acoustics Conference (AFPAC 2016)
13–15 January 2016, Surrey, United Kingdom

Accepted papers received: 20 December 2016Published online: 18 January 2017

Preface

Papers

15th Anglo-French Physical Acoustics Conference (AFPAC2016)

FOREWORD/PREFACE

The 15th Anglo-French Physical Acoustics Conference (AFPAC) was held at Selsdon Park Hotel, near London, United Kingdom, on 13-15 January 2016. The venue was an excellent location to exchange ideas, regardless whether this took place in the conference room, over lunch, at the drinks reception, or in the bar after the conference dinner. A total of 65 papers were presented at the conference. There were over 80 delegates from institutions covering five countries.

On the first day of AFPAC, the Institute of Physics joined forces with the National Physical Laboratory (UK) to host a special session on cavitation. The Cavitation User Forum, a bi-annual event specifically dedicated to applications of high power ultrasound, brought together experts from academia and from the cleaning, processing and medical industries. This session was kicked off with an invited talk by Dr David Fernandez Rivas (University of Twente, The Netherlands), on the reproducibility of sonochemistry and ultrasonic cleaning.

The Cavitation User Forum was followed by a special session on biomedical ultrasound, co-sponsored by the Medical Physics Group of the Institute of Physics, which featured a keynote talk by Prof Robin Cleveland (University of Oxford) on ultrasonic surgery. The session included talks on acoustic microscopy of live cells, histotripsy, phase-insensitive ultrasound computed tomography for the diagnosis of breast cancer, high-intensity focused ultrasound and the biomedical applications of solitary wave impulses generated by granular chains

The second day featured an invited presentation by Prof Tim Leighton (University of Southampton, UK) on the acoustic bubble, which discussed ocean, cetacean and extra-terrestrial acoustics, and cold water cleaning. Prof Christ Glorieux (KU Leuven, Belgium) discussed the applications of photothermal and photoacoustic methods using different spatiotemporal excitation patterns. A broad range of physical acoustics topics was reviewed that day. Work was presented on phononic crystals, high resolution imaging of in-depth rough interfaces, shear wave reconversion in nano-fluids and the prediction of radiation by finite-sized sources in semi-infinite isotropic plates. The field of non-destructive evaluation was well represented, with a presence from institutions such as Imperial College London, University of Nottingham and Université de Montpellier.

During the Conference Dinner, the 2015 Bob Chivers prize was awarded to Dr Bo Lan for his paper entitled "Experimental and computational studies of ultrasound wave propagation in hexagonal close-packed polycrystals for texture detection".

IOP Conf. Series: Journal of Physics: Conf. Series **797** (2017) 011001 doi:10.1088/1742-6596/797/1/011001

On the final day of the conference, Prof Olivier Dazel (Université du Maine, Le Mans, France) gave an invited talk discussing the past present and future of the modelling of sound absorbing materials.

The UK organising committee was particularly happy to welcome the many French contributors that travelled to Selsdon Park and would like to thank Alain Lhémery and the Société Française d'Acoustique (SFA) for publicising the event in France. We are happy to announce that many of the presented papers will be published in the conference proceedings in Journal of Physics: Conference Series. We would also like to draw the reader's attention to the upcoming 16th AFPAC conference that is scheduled to take place on 23-25 of January 2017 at the Laboratoire de Mécanique et d'Acoustique, Marseille, France.

Pierre Gélat (UCL, United Kingdom), Theodosia Stratoudaki (University of Nottingham, United Kingdom), Rachel Edwards (University of Warwick, United Kingdom), Robert Malkin (University of Bristol), Nader Saffari (UCL, United Kingdom) and Alain Lhémery (CEA, France)

Peer review statement

All papers published in this volume of *Journal of Physics: Conference Series* have been peer reviewed through processes administered by the proceedings Editors. Reviews were conducted by expert referees to the professional and scientific standards expected of a proceedings journal published by IOP Publishing.

The acoustic bubble: Ocean, cetacean and extraterrestrial acoustics, and cold water cleaning

T G Leighton

Institute of Sound and Vibration Research, Faculty of Engineering and the Environment, University of Southampton, Highfield, Southampton SO17 1BJ, UK

Abstract. This paper summarizes the content of a plenary lecture on the author's personal research into the interactions between bubbles and sound fields, covering particular topics involving the climatically important gas exchange between atmosphere and ocean, the implications of bubbly ocean water to marine mammals that use sound, and the opportunities afforded by incorporating acoustical sensors onto probes launched to investigate other worlds in our solar system. It closes with recent data on the opportunities of bubble acoustics to investigate methods of cold water cleaning.

1. Introduction

Gas bubbles in liquids have an extraordinary ability to interact with sound fields. Although traditionally many people date the start of studies on collapse cavitation to the 1917 work of Rayleigh [1], in fact Rayleigh's well known analysis was predated by some 70 years by Stokes' handwritten solution to an examination question on the problem that he set in 1847 for less able physics undergraduates (see Ref. [2] for details). Those analyses were dominated by the effect of the inertia of the inrushing liquid as it collapsed towards the centre of the void. Noltingk and Neppias [3,4] examined how such a cavity could be created and so introduced some permanent gas into the void, although the dynamics of the collapse were still dominated by the liquid inertia. The introduction of permanent gas harmonized the description of such 'inertial' cavitation with the famous 1933 description of non-inertial cavitation by Minnaert [5], where the effects of liquid inertia and gas stiffness are balanced to determine the natural frequency of oscillation of a pulsating bubble.

As Minnaert himself hypothesized, bubbles generate the song of a babbling brook [5,6], and ocean sounds through breaking waves. Such ocean sounds, and the way the oceans scatter and absorb sound, have since helped us understand the global carbon budget, as this paper will outline, Section 3 will describe how the particularly dense clouds of bubbles are produced in the ocean by humpback whales to form 'bubble nets' to catch prey, and speculate on the acoustical implications for this. The acoustical implications are wholly different when dolphins hunt with bubble nets, as they must adapt their sonar to avoid the bubbles themselves preventing them from finding their prey using echolocation.

Section 4 will explain how a desire to predict the sounds of Titan prior to the first probe landing there, led to a wider exploration of the numerous ways that we might use and observe sound on other worlds. Particular attention has been paid to improving the understanding of extraterrestrial acoustics so that our familiarity with sound on Earth does not lead us into assumptions that introduce errors into the design of acoustic sensors for probes, or produce erroneous interpretations when we eventually collect acoustic signals from other worlds.

IOP Conf. Series: Journal of Physics: Conf. Series **797** (2017) 012001 doi:10.1088/1742-6596/797/1/012001

To conclude the paper, Section 5 describes a new application of bubble acoustics, namely cold water cleaning without additives, and discusses its possible application in a range of scenarios where the use of cold water is preferable or unavoidable.

2. Bubble detection in the natural world

Figure 1 shows the simultaneous hydrophone record [panel (a)] and high speed video [panels (b) to (h)] recorded when a water drop impacts upon a body of water. This process is familiar for images resembling those in the final frame [panel (h)], where the half-submerged lens clearly reveals the water jet that rises into the air as, below it, the crater shrinks in the water surface. This image is associated with the famous 'plink' of a dripping tap. That 'plink' can be seen as the exponentially decaying sinusoid in the hydrophone trace of panel (a), the labelling indicating that it is synchronous with panel (f). Therefore the famous 'plink' sound is not caused by either the jet or the crater, but by the tiny bubble that was pinched off from the base of the closing crater [panels (e) to (g)].

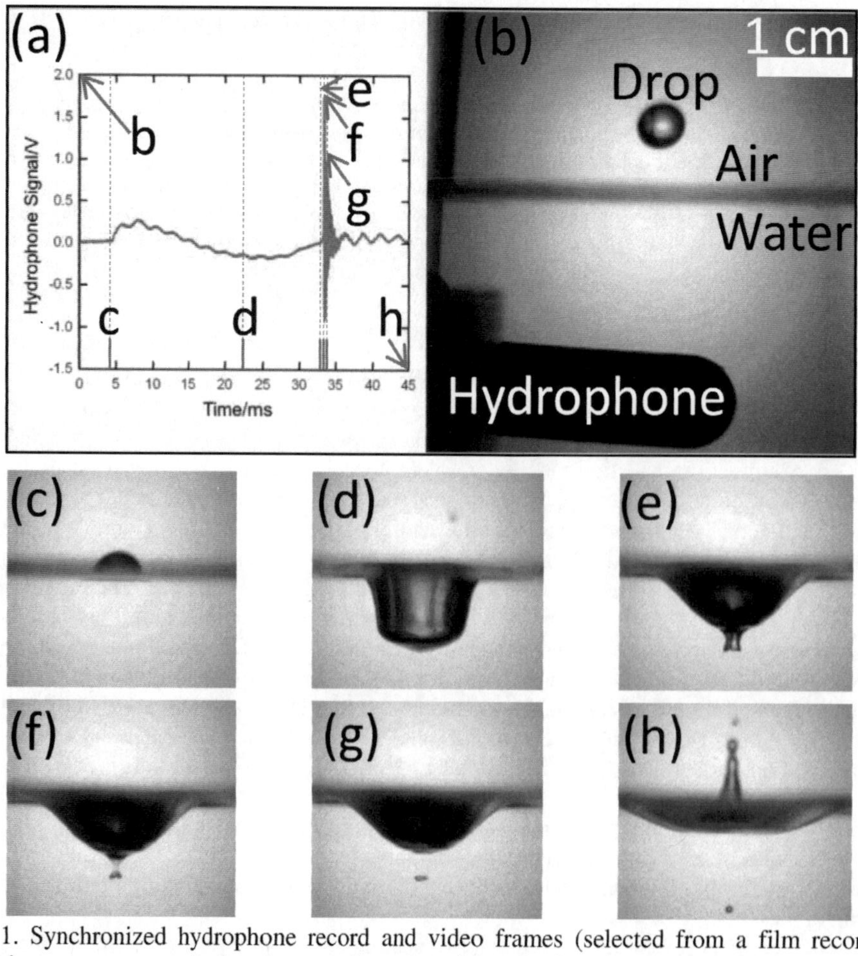

Figure 1. Synchronized hydrophone record and video frames (selected from a film recorded at 10,000 frames per second) for the sound produced by the impact of a droplet on a water-air interface. (a) The hydrophone output (in red, with the timing of subsequent frames labelled (b)-(h) showing the impact of a water drop falling from air into water. The hydrophone signal from (c) to (d) is hydrodynamic, the only significant acoustic emission occurring when a small bubble is pinched off from the base of the crater ((e)-(g)).

This experiment illustrates how powerful an acoustical entity is a gas bubble in liquid. Each bubble behaves like an underwater bell, small ones producing plinks of high notes, and larger ones generating low notes [5]. Therefore from the pitch of the 'plink', one can determine the size of the bubble. The first count of the size distribution of bubbles [6] entrained in the natural world, made using the sounds they generated, was undertaken in the early 1980s. This led to similar counts for the bubbles trapped by rainfall over the ocean [7,8], and today we see the deployment of at-sea acoustic monitors for rainfall [9]. The technique was also deployed in the ocean [2,10,11] to detect bubbles trapped under breaking sea waves. When an ocean wave breaks, it generates many bubbles, each 'singing' its own note, and from the overall sound we can determine the number and size of bubbles containing trapped atmospheric gas, which can form clouds [12,13] in the upper ocean. These bubbles are responsible for the transfer between atmosphere and ocean of many hundreds of millions of tonnes of atmospheric carbon each year.

However, to quantify this climatically important carbon transfer, it is not sufficient simply to know how many bubbles are injected into the ocean by breaking waves. One must also know how many are left some time after the wave has broken, after some bubbles have risen to the surface, and others have dissolved. To do this, we developed techniques to measure the 'silent' bubbles whose ringing ceased some time ago, based on projecting sound at the bubble and re-exciting them to emit sound. One particularly useful discovery [14] was that, when a signal with a high frequency (f_i) is projected at a bubble cloud at the same time as a signal at a lower frequency (f_p), then the bubbles that are resonant at f_p can uniquely scatter the frequency $f_i \pm f_p / 2$, allowing bubbles of this size to be identified from clouds of other bubbles [15-19]. By varying f_p , a cloud of bubbles could be scanned to count and size them all uniquely [20]. This was used to count bubbles in the sea, and as the basis for the development of a range of techniques suited to oceanic bubble counting [20-22], particularly in the surf zone where previous acoustical methods had lacked the ability to cope with the time dependent and nonlinear [22,23] effects that would occur there.

3. Cetacean acoustics

Our ability to model the scattering of sound by undersea bubbles allowed us to postulate the mechanism by which humpback whales trap prey within spiral bubble nets. Although it had been known for decades that whales blow bubbles to do this, the reason why the prey do not escape the trap was not known [2]. Our models (Figure 2(a,b)) showed that the spiral bubble net traps the loud calls emitted by whales to produce an impassable 'wall of sound', whilst simultaneously creating a quiet zone in which the prey would congregate, this zone occurring in the model at the location where the rising whales feed (as photographed in Figure 2(c)) [24].

However, unlike humpback whales, dolphins use high frequency sonar to find prey, and the bubble nets they create (Figure 2(d)) would confound their sonar. Rather than accept that dolphins would 'blind' their most spectacular sensory apparatus when hunting, we set about proving that a previously unknown type of sonar processing (TWIPS) could detect prey in bubble nets [25, 26], and showed this to work with dolphin sonar calls [27]. Although the question of whether odontocetes use such a method or not is still open to question [28, 29]. Industry is now developing this to protect shipping in coastal regions such as the Persian Gulf, where clouds of bubbles and particles in the near-shore waters make mine detection difficult (Figure 3(a)). Realizing that this new processing system could work with radiations other than just sonar, we used it to develop a radar system (TWIPR) where the scattering off circuitry from a bomb trigger was in excess of 30 dB more powerful than the scattering off other targets (Figure 3(b)) [30]. With the ability to detect mobile phones as readily and selectively as bomb triggers, TWIPR can help to find buried targets of interest (bombs, people carrying phones buried by collapsed buildings or avalanches) where normal radar would not be able to identify the genuine target from other debris (typified in Fig 3(b) by (i) & (iii)).

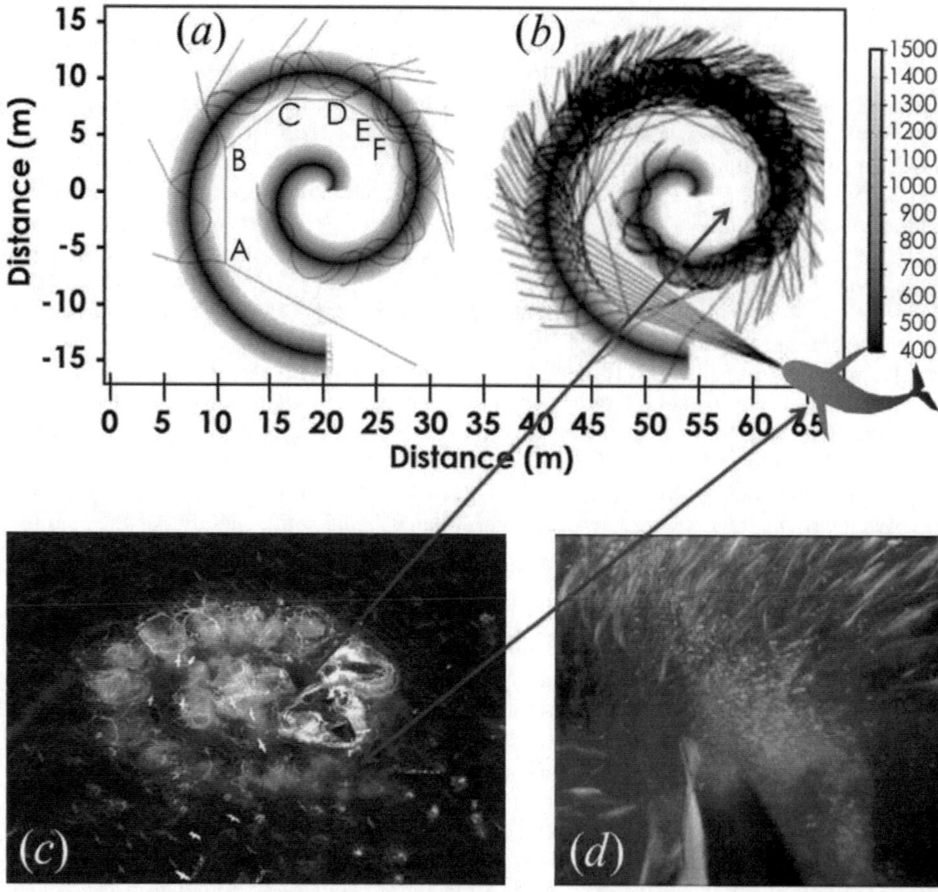

Figure 2. Modelling the acoustics of whales and dolphins when they use bubble nets for hunting. (a) A plan view 2D map models the bubble-induced variation in sound speed (shown in greyscale) that has been inferred from photographic evidence of spiral bubble nets produced by humpback whales. A single ray, projected into the open end of the spiral, impacts the bubbly wall in turn at A, B, C, D…, the grazing angle decreasing each time such that it does not propagate all the way to the centre of the spiral. At each impact with the bubble wall, the ray is partially reflected from the bubbly wall and partially transmitted into it, from which point refraction occurs. No absorption is included. (b) When many rays are modelled in this way, a 'quiet' zone is seen to occur at the centre of the net, walled-in by zones of high intensity. The photograph in (c) shows that this corresponds to the point at which whales rise from the water (purpose arrow connecting the locations of the source modelled whale and quiet zone) (photo by T. Voorheis of Gulf of Maine Production). (b) Image of a dolphin blowing bubbles to catch fish (Image courtesy of The Blue Planet, BBC). For details see Refs. [2, 31-33].

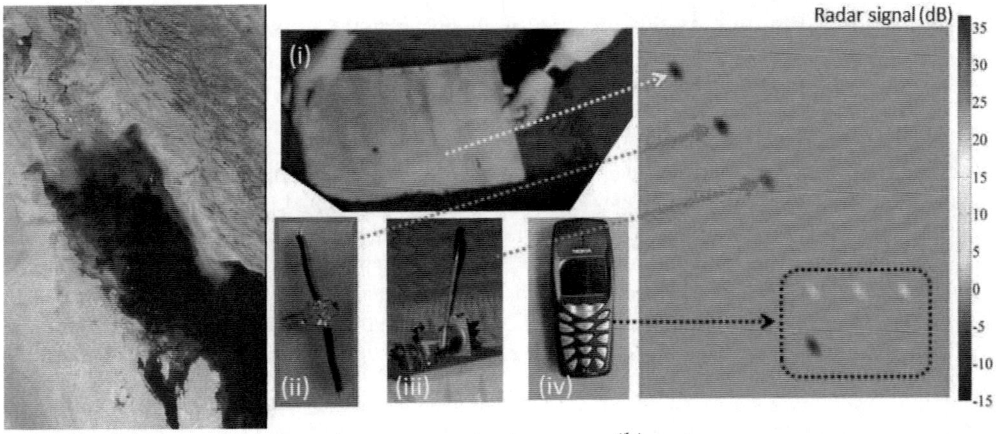

(a) *(b)*

Figure 3. Scenarios where the use of two pulses to distinguish target from clutter would help (a) sonar and (b) radar. (a) Aerial image of Persian Gulf (image courtesy J. Descloitres, MODIS Land Rapid Response Team at NASA GSFC). (b) TWIPR Radar signal from (i) an aluminium plate, (ii) a circuit resembling components of a bomb trigger; (iii) a rusty bench clamp; (iv) mobile phones in various states: on, off, or with invalid SIM cards, which (along with differences in the radar resonance characteristic of the semiconductors with different models of phones contained within the dotted box) gives rise to different scattering strengths. The spatial layout of the colour map reflect the superimposition of different tests, because the test rig was not sufficiently large to lay out all targets simultaneously. See Ref. [30] for details.

4. Extraterrestrial acoustics

The ability to infer the bubble sizes generated from the sounds of waterfalls, breaking waves, and rainfall (as discussed in Section 2) was used to create [2, 34] the possible sound of 'methanefalls' (waterfalls made up of liquid methane and ethane) on Saturn's largest moon, Titan. As the Cassini-Huygens mission approached Titan in 2004, no-one knew what the surface would be like because Titan is shrouded in a thick fog (Figure 4(a)). However, one body of opinion held that, with a 93 K surface temperature, the cold conditions and dense atmosphere would allow for the existence of lakes and possibly methanefalls on Titan (Figure 4(b)).

Prior to Huygens' landing, we simulated the sound that would be made were Huygens to splashdown in a lake, and the sound that a probe on the surface of Titan might detect if it landed with its camera facing away from the methanefall. Huygens was very successful, and although its images from its landing site revealed a barren landscape (Figure 4(c)), during descent some indication of topography that might have been carved by flowing surface liquid was revealed (Figure 4(d)), and later radar observations by Cassini revealed lakes (Figure 4(e)).

The objective of our research was to provide material for outreach, but also to explore the extent to which we might start to construct the soundscapes of other worlds. Despite all the planetary probes that have been sent out, we have not yet heard the soundscape of another world [35]. Consequently, this work was conducted for the purpose of:

- enabling better design of microphones and sound sources for use on future planetary probes, with respect to improving signal to noise ratios, more reliably interpreting any detected signals, and enabling and maintaining an appropriate calibration given the differences between Earth's environmental parameters and the deployment environments (which can vary hugely even over a single world like Venus or Jupiter) [36-40]);
- improving the design of missions exploiting acoustics in planetary exploration (for

IOP Conf. Series: Journal of Physics: Conf. Series **797** (2017) 012001 doi:10.1088/1742-6596/797/1/012001

example, by correcting the analysis used to predict the correct placement of detectors on ice-covered moons like Jupiter's moon Europa, with the purpose of using sound to explore the vast water oceans beneath the ice [41-43]);

- eliminating errors in mission planning introduced by use of familiar Earth-based acoustics to extraterrestrial environments [38];
- exploring the extent to which we might interpret sounds picked up by planetary probes to ascertain key features about the world the probe is exploring [44,45] (Figure 5).

For this latter objective, in designing the algorithms to simulate the sounds of worlds, we were able to provide a device, licensed to planetaria [44], which not only allows the audience to hear the simulated sound of the world under discussion, but in live presentations allows the presenter to use the voice they would have on a given planet (if they could live and speak), when telling schoolchildren about that planet [46,47]. More details on this topic are available on the Internet [48].

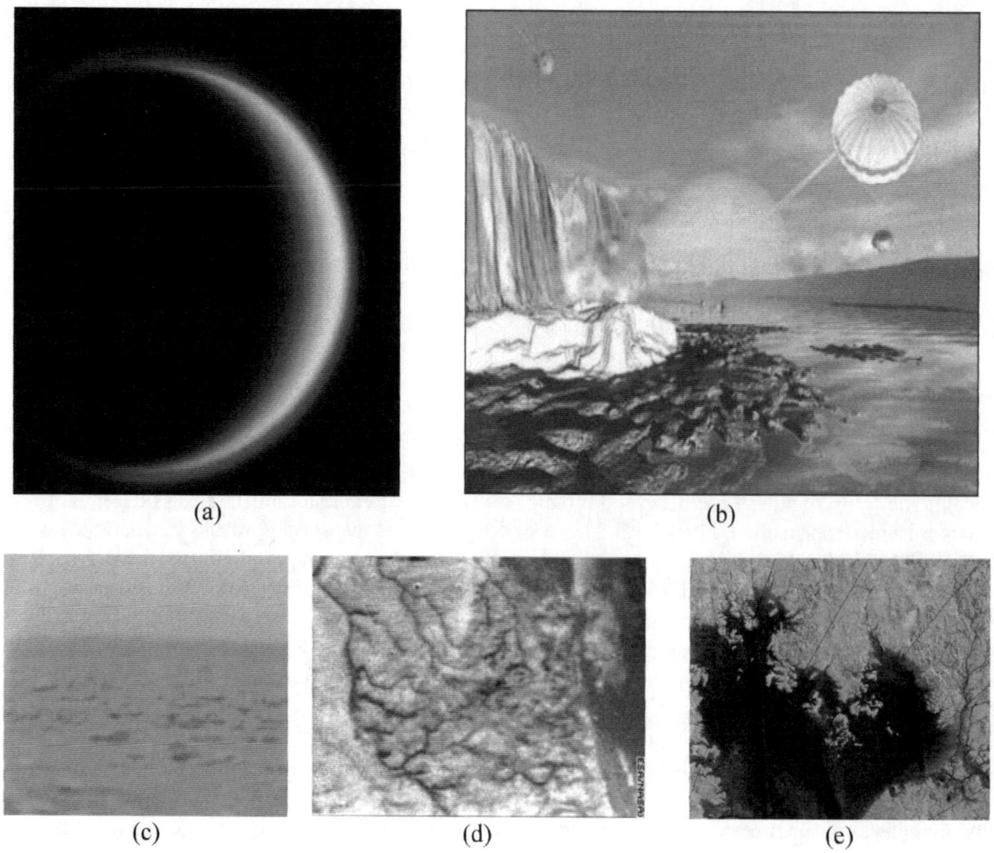

Figure 4. Images of Titan of relevance to the *Cassini-Huygens* mission. (a) Image of Titan. (b) Artist's impression of the *Huygens* probe parachuting through Titan's atmosphere, having previously detached from the *Cassini* vehicle (seen in the upper left of the image). (Painting by D. Seal). (c) The surface of Titan as imaged by the Huygens probe after landing. (d) Images of the surface of Titan taken by *Huygens* during descent. (e) False-colour Cassini radar image of Titan's surface. Blue colouring indicates low radar reflectivity, attributed to hydrocarbon seas, lakes and tributary networks filled with liquid ethane, methane and dissolved nitrogen. All image credits: NASA/JPL/Caltech.

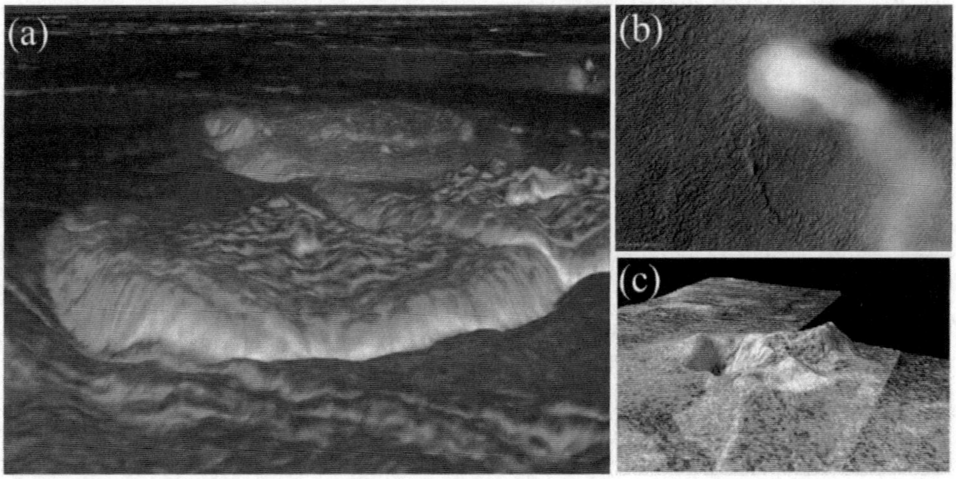

Figure 5. Extraterrestrial locations of interest for acoustical sensors. (a) A computer reconstruction of the surface of Venus, created from data from the Magellan spacecraft. Credit: E. De Jong et al. (JPL), MIPL, Magellan Team, NASA http://www.space.com/18525-venus-composition.html (b) Dust devil recorded by the HiRISE camera on board the Mars Reconnaissance Orbiter. Tracking across the flat, dust-covered Amazonis Planitia in the northern Martian spring of 2012, the core was about 140 meters in diameter. Lofting dust into the thin Martian atmosphere, its plume reaches about 20 kilometers above the surface. Tangential wind speeds of up to 110 kilometers per hour are reported for dust devils in other HiRISE images. Image Credit: HiRISE, MRO, LPL (U. Arizona), NASA.http://apod.nasa.gov/apod/ap150303.html. (c) False colour image of evidence for a possible cryo-volcano on Titan, photo credit, NASA, Cook *et al.* [49].

5. Cold water cleaning

4.1. The need for cold water cleaning

Efficient cleaning is central to modern life, from healthcare, to food preparation, to manufacturing and maintaining infrastructure [50,51]. This not only constitutes excessive consumption of one of our most valuable resources, but also generates large quantities of contaminated run-off. Uses range from the domestic to the exotic. Managing and purifying the water used in cleaning is a major issue because:

- the volumes of water used in cleaning are very great (e.g. it takes 100 tonnes of water to produce 1 tonne of clean wool after shearing [51]);
- purifying run-off is costly (e.g. each cubic meter of water used for cleaning in the nuclear industry costs £10,000 to treat subsequently [51]);
- even water that might appear not to be as grossly contaminated as the above two examples can present a hazard because of the volumes are high and run-off tends to travel into natural water resources (with detergents poisons, pesticides and herbicides from farmland and industry being carried into streams [52] where they can cause the demise of fish populations [53]).

Policymakers, the public and industry need to act to future-proof society against future problems caused by the extensive use of water in cleaning. Already, inadequate cleaning produces severe public health issues in hospitals [54, 55]. From major factories in the industrialized world to the local abbatoir in the developing world, the mismatch between the volumes of water and additives used, and the robustness of the water supplies on which we depend, is failing to do its part in mitigating against future crises in water supplies, food production, and healthcare [56].

4.2. The Ultrasonically Activated Stream

One particular innovation was the Ultrasonically Activated Stream (UAS, invented at the University of Southampton and now in production by Ultrawave Ltd. under the name StarStream^{TM}), which enhances the cleaning ability of liquids, and in particular enables cold water cleaning. The UAS system, like the ultrasonic cleaning bath, bases its cleaning action on the speed of the bubble wall motion, and not (like pressure washers) on the speed of the flow. As such it will be less damaging. The UAS system produces a free flowing liquid stream, each nozzle generating flows of around 1-2 litres min^{-1}. Here the cleaning action of bubbles, excited with a suitable ultrasonic field, is generated at the end of a fluid stream. In addition, low flow rates of fluid within this approach are useful in releasing the contaminant from the surface and avoiding re-deposition at another location (a further possible limitation in bath geometries). Using higher frequencies and generating lower amplitudes of ultrasound in air than cleaning baths, StarStream has not had any reports of the 'subjective' adverse effects (headaches, nausea, tinnitus, migraine etc.) anecdotally attributed to some other sources of ultrasound in air, including some (but by no means the majority of) users of ultrasonic cleaning baths [57]. The low velocity stream approach has many advantages; however, two basic criteria are necessary for this strategy to be successful. First, the sound field must be sufficient to generate bubble activity at the solid/liquid interface of the material to be cleaned. Second, a suitable bubble population must also be present on the surface of the target that needs cleaning. This population can then be driven by the sound field deployed and act on the contaminant at the interface in question (through suitable oscillation [58-61] and shear forces). These two requirements are by no means trivial to create within a flowing stream but this has successfully been achieved in the UAS system.

The UAS system in some ways brings the power of an ultrasonic cleaning bath to the end of a hose, so that items can be 'cleaned in place'. Both the UAS and the ultrasonic cleaning bath replace the pressure and flow that comes purely from the stream of water in the pressure washer, with pressure and flow close to a bubble wall. However, it would not be correct to equate the bubble activity in the cleaning bath with that which occurs in the UAS. The ultrasonic cleaning bath causes cavitation, whereby bubbles collapse under pressure fields to generate extreme conditions [62-64], which can include the generation of free radical generation [65, 66] and strong pressure waves [67-68] from the gas compression, although significantly more powerful ones can occur in the blast waves launched when collapsing bubbles involute to form microjets [69-72]. Pressure waves and jet impact can remove material from surfaces [71, 73]. In contrast, the UAS system projects sound down a column of water [74] in order to excite surface waves [75] on the walls of microscopic bubbles on the surface to be cleaned. These surface waves can generate convection [76-79] and shear forces [80, 81] in the liquid close to the bubble wall, and so produce a cleaning effect, and alter the way material deposits onto surfaces [82].

The design, construction and operation of the UAS device are detailed elsewhere [83], but the basic principle is that cold water is fed into a hollow horn that contains an ultrasonic transducer operating in excess of 100 kHz. The ultrasound and microbubbles in the flow both travel down the stream of water to the target that is to be cleaned. If the bubbles are ultrasonically activated when they are on the target, the cleaning ability of the liquid is enhanced in four ways (figure 6):

- The bubbles are attracted to the surface to be cleaned by Bjerknes radiation forces [84], and are not as rapidly washed away by the flow as they would be in the absence of ultrasound.
- The bubbles are particularly attracted into crevices by secondary Bjerknes radiation forces [84]; such crevices are traditionally more difficult to clean by wiping or brushing.
- Surface waves on the walls of the bubble, excited by the ultrasound, produce enhanced convection in the liquid and enhanced shear in the contaminant, causing its removal.
- The progress of the bubble into the crevices would, if the liquid contained additives (e.g. detergent or biocide) cause that additive to penetrate the crevice far more rapidly than would reliance on simple diffusion, so that cleaning can potentially be achieved more rapidly, and with lower concentrations of additives.

IOP Conf. Series: Journal of Physics: Conf. Series **797** (2017) 012001 doi:10.1088/1742-6596/797/1/012001

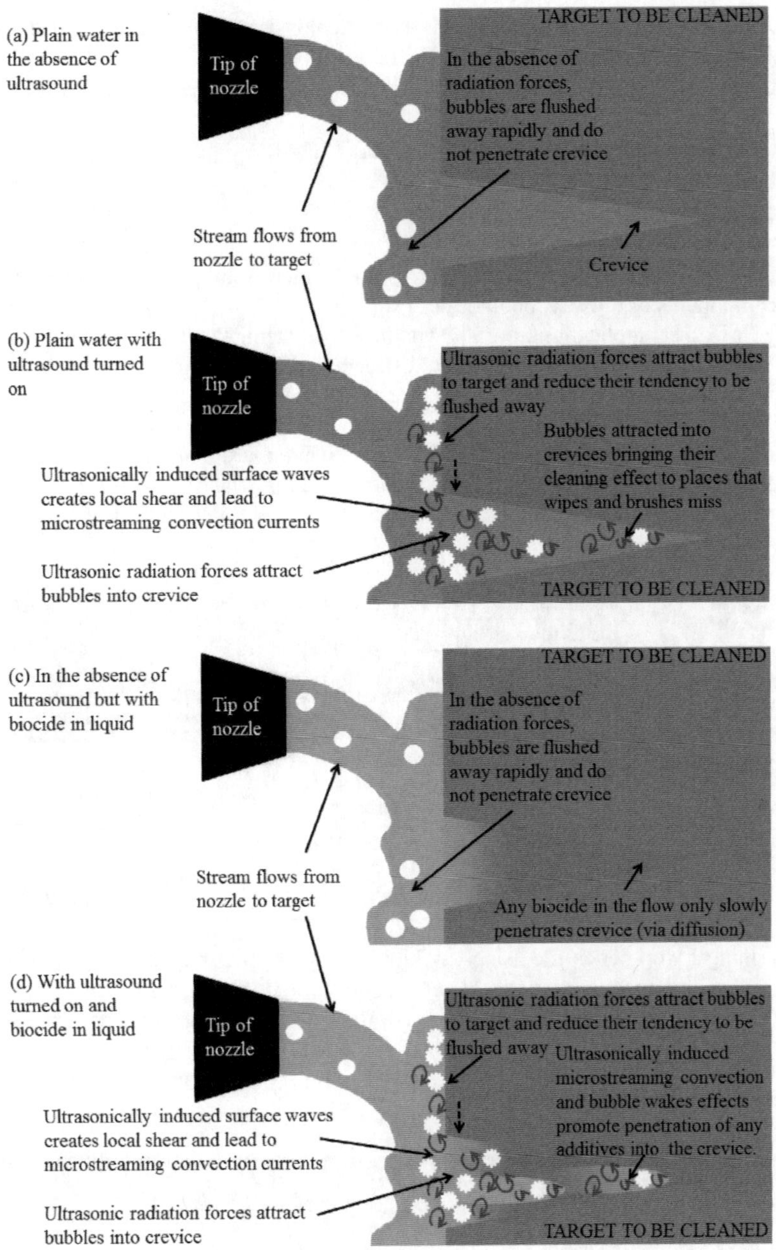

Figure 6. Schematic comparing the behavior of bubbles emitted from a StarStream device, for cases when: (a) no ultrasound is used, and the water contains no additives; (b) the water contains no additives, but the ultrasound is activated; (c) no ultrasound is used, and the water contains additives (in this case, biocide); (d) the water contains additives (in this case, biocide) and ultrasound.

The effectiveness of the UAS system has been demonstrated in controlled tests on particularly difficult problems, specifically on:

- the cleaning of brain tissue and prions from surgical steel, the removal of contaminating material from bone transplants [85];
- the removal of biofilms of dental bacteria [85, 86];
- the cleaning of skin models [85, 87];
- the cleaning of marine biofoulant [88];
- the cleaning of railway track [89];
- the cleaning of hands, kitchen surfaces, tools, glue from jar labels, contaminated tubes, and components of railway locomotives [90].

Here a range of other applications are shown. In all of them, the UAS system was a commercial StarStream device that projected, into the liquid stream, only cold water directly from the mains water supply, without additives or heating, at a flow rate of around 1 litres per minute.

Figure 7 shows how StarStream enables cold water, without additives, to remove mascara makeup from a metalworker's file. Figure 8 shows removal of whiteboard marker from a hand by StarStream using just cold water with no additives. Figure 9 shows the removal of a layer of Vaseline™ petroleum jelly (beneath which was a layer of lipstick, added to provide a visual cue for the removal of the Vaseline™).

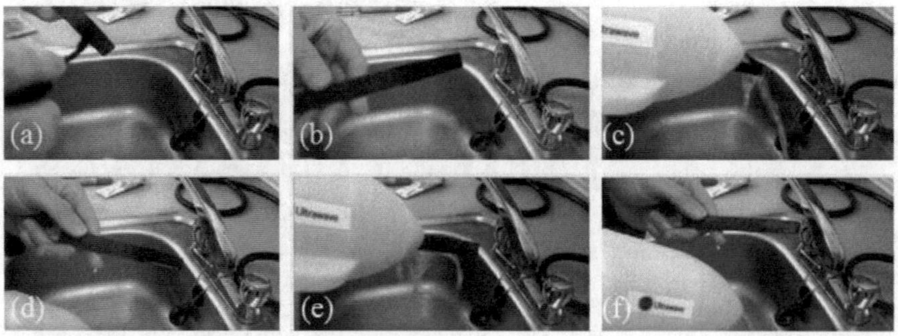

Figure 7. Cleaning a workbench file using UAS. (a,b) A workbench file has mascara make-up smeared onto it. (c) A StarStream device – with the ultrasound turned off - passes a stream of cold water with no additives over this contaminant, and afterwards (d) there has been no observable cleaning. However, when (e) the same water stream has ultrasound added to it, then afterwards (f) both the contaminant has been removed. The panels are stills from the video available at: https://www.youtube.com/watch?v=22tUGEGyk10

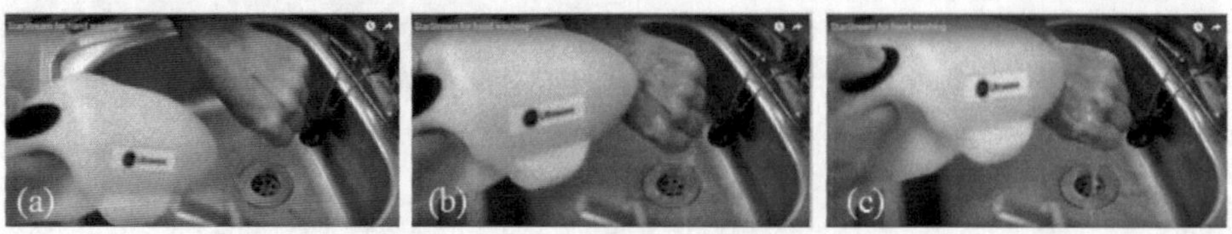

Figure 8. Cleaning whiteboard marker off a hand using UAS. (a) A line of red whiteboard marker is applied to a hand, but (b) when a StarStream device – with the ultrasound turned off - passes a stream of cold water with no additives over this contaminant, no observable cleaning occurs. However, when (c) the same water stream has ultrasound added to it, then the hand is cleaned. The panels are stills from the video available at: https://www.youtube.com/watch?v=ElxcBn7-t8s

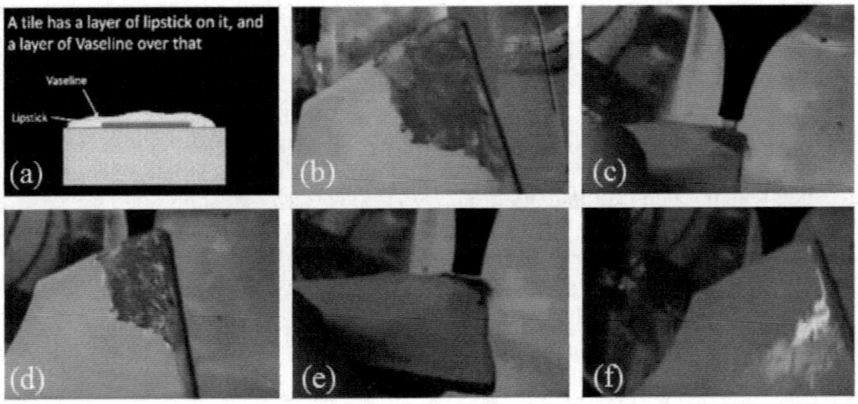

Figure 9. Cleaning lipstick, covered by a protective layer of Vaseline™, using UA. (a,b) A ceramic tile (10 cm x 10 cm) has lipstick placed on one corner, and a layer of Vaseline™ petroleum jelly placed over the lipstick. (c) A StarStream device – with the ultrasound turned off - passes a stream of cold water with no additives over this double contaminant, and afterwards (d) there has been no observable cleaning. However, when (e) the same water stream has ultrasound added to it, then afterwards (f) both layers of contaminant have been removed. The panels are stills from the video available at: https://www.youtube.com/watch?v=XZ2oe4XaM3Y

4.2. Choice of device

There are many options for cleaning. Perhaps the most widespread for external use are domestic and small-industry power washers (also known as pressure washers) which pass water (usually with additives) at high speed onto the surface to be cleaned. Contaminants are flushed away (so that unlike systems that clean immersed targets, the object to be cleaned is not immersed in a 'soup' of contaminated liquid). The down-side to this is that power washers use large volumes of liquid (up to 20 litres per minute for a large pressure washer), and consequently lead to large volumes of contaminated run-off. Pressure washers are effective, and if resources are significant (for example, on a warship or on a train) they can always be scaled up to produce enormous pressure. As a result, in assessing the performance of a pressure washer against any alternative, the first question is not 'which cleans best' but rather 'can I tolerate the pressures, splashback, aerosol, run-off and potential damage that the candidate pressure washer might produce?'.

On the domestic or small-business scale, pressure washers are difficult to scale up (in terms of using multiple nozzles or larger nozzles), both because of the pump requirements (in terms of power and water usage), and because they generate considerable back force when used, meaning that the structural support for scaled up versions must become increasingly robust. Indeed, even on major infrastructure like the railways, although pressure washers can be used to try to mitigate the effect of 'leaves on the line', those washers are so powerful that they damage the track if used when the locomotive is stationary, and when in normal use (with the locomotive moving) they can damage the locomotive from the stones that the pressure washers accidentally throw up, and the run-off can damage the banks on which the track sits. This challenge is discussed in more detail later.

Even small power washers can be damaging to delicate surfaces (a well-recognized mistake by boat owners who attempt to clean rope with pressure washers), and they generate aerosols of contaminated liquid (the liquid droplets produced when pressure-washing, say, sewage lines represent a significant hazard and route to spread contamination [51]). Whether the user cares about the backsplash, aerosol and spray generated by pressure washers, and the abilities of these to carry and redistribute contaminants onto nearby objects including the user (where they can be inhaled, settle on skin or eyes etc.), depends on the application. Applications where such phenomena might be of particular concern include cases where the contaminant contains sewage, bacteria, radionucleotides, petrochemicals or chemical ingredients that represent an environmental hazard (e.g. marine antifoulant, microbeads etc.).

Pressure washers will entrain an *ad hoc* population of bubbles. There are many technologies advertised as enhancing cleaning by the addition of bubbles, often with size distributions constrained to certain small sizes [91-94] but without going to the trouble (in terms of instrumentation, power requirements etc.) of adding sound, begging the question of why one would ever wish to use an ultrasonic system, with its added complexity of needing to power an ultrasonic source.

The ultrasonic field does two things: it causes motion of the bubble wall (e.g. generating inertial cavitation in the ultrasonic cleaning baths, and surface waves with UAS systems); and it attracts the bubbles towards the solid and into crevices. Figure 10 compares the Primary Bjerknes force that attracts this bubble to the rigid wall, with the Mutual Bjerknes force which would be attractive between two bubbles whose centres are separated by the distance given on the horizontal axis (an unrealistic proposition for distance smaller than a bubble diameter). As expected [95, 96], the Primary Bjerknes force that attracts the bubble to the wall is many orders of magnitude greater than the other two forces: although it must in principle decrease to zero at the bubble wall, of course these numbers lose meaning within one bubble radius (here, 20 microns) from the wall. Coincidentally, this is roughly the distance from the wall at which the Mutual Bjerknes force (which in principle varies with distance from the wall as an inverse square law [96]) equals the buoyancy force on the bubble.

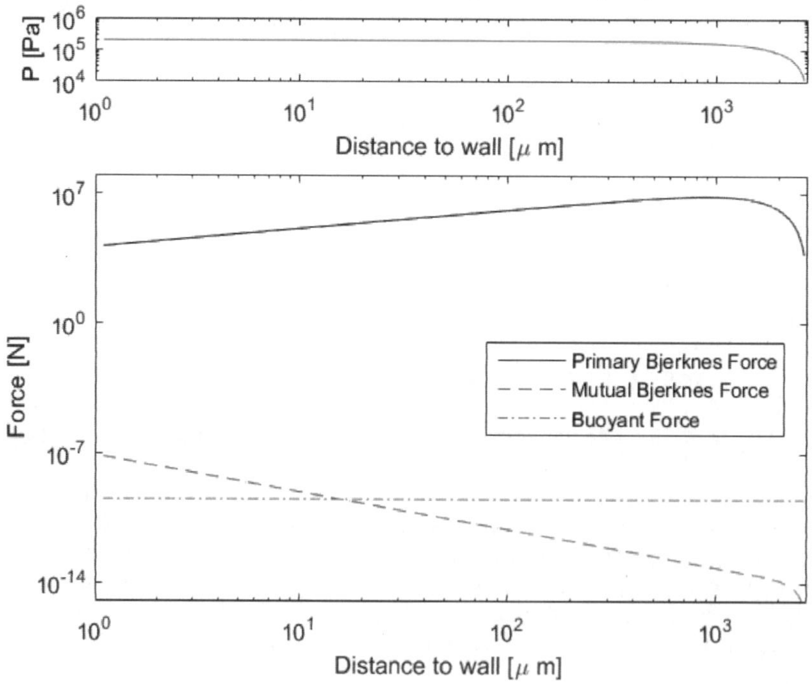

Figure 10. Comparison of pressures and forces acting on a bubble near a wall, plotted on a common horizontal axis that shows distance from a rigid wall (limited to within one quarter wavelength from the rigid wall). The upper panel shows the variation of the amplitude of the time-varying pressure, formed when a 135 kHz plane wave of amplitude 100 kPa is normally incident on the rigid wall. The lower panel compares the Primary and Mutual Bjerknes forces on an air bubble in water under 1 bar static pressure. The buoyancy force is shown for comparison. Note that both axes are logarithmic. The calculation assumes simple harmonic motion of the component of bubble motion that contributes to the zeroth order spherical harmonic perturbation of the bubble wall (in practice such high driving pressures will cause departures from this in this bubble).

The magnitude of the Primary Bjerknes force that attracts the bubble to the wall is impressively large, given that the water hammer pressure exerted by the flow of the liquid stream would be only 20-40 Pa, 100,000 less than the water hammer pressure from a domestic pressure washer. A pressure washer would of course not be capable of using that water hammer pressure as an active force to drive a bubble to a wall, as the Primary Bjerknes force does (and of course the acoustic radiation forces that attract bubbles to surfaces will increase if the wall is structured [97]). Therefore the radiation forces that attract the bubble to the wall are impressively high, but it is critically important to recall that these are the forces causing bubble migration; they must be distinguished from the forces that cause cleaning. The forces associated with the actual cleaning are then the shear forces close to the bubble wall produced by surface waves.

Addressing therefore the use of ultrasonically activated bubble activity in cleaning, the most well established device for cleaning using bubbles and adding sound, is the ultrasonic cleaning bath [98], which most frequently relies on an *ad hoc* bubble population. As an immersion system, it does not generate large quantities of run-off and can save time by bath-processing many units simultaneously. Such baths represent an established global product, but have many limitations which prevent further market penetration: the bath is too small for some objects (e.g. they cannot be used for vehicle cleaning); the bath cannot get into complex geometries and clean them (e.g. it cannot be used to hose down inside an engine); the item to be cleaned sits in a soup of contaminated liquid. Moreover, immersion of the object to be cleaned into the bath can degrade the sound field and the cleaning action [51]. This is not always apparent, since the monitoring of the cleaning performance of such ultrasonic baths is often rudimentary, much of industry eschewing new techniques in favour of a check based on the insertion of domestic aluminium cooking foil into the bath to see whether the cavitation is capable of generating small erosion pits and holes within the foil [98, 99]. The above list of potential drawbacks to ultrasonic cleaning baths is not exhaustive when considering a procedure such as hand washing: the ultrasonic bath works by generating violent bubble collapse, which may be destructive to tissue such as skin, and can penetrate tissue at bath frequencies.

Comparison of UAS to ultrasonic cleaning baths and pressure washers is illuminating, in that it shows that the question is less of how well a technology cleans, but more of in what scenario it can be used. Ultrasonic cleaning baths and UAS both use ultrasonically activated bubbles, but the mechanism by which they clean is very different, and this produces key observables. For example, one would never immerse a hand in an ultrasonic cleaning bath because the hazard of cell damage is high, and yet UAS has to date been shown to be safely used on hands (figure 8). One might therefore conclude that UAS uses a 'gentler' form of cavitation, and few would argue that Faraday waves represent a 'gentler' form of bubble activity than the inertial cavitation that the bath generates. However having settled on the label of 'gentler' cavitation, one must not confuse gentleness with effectiveness: in a recent study [89, 100], UAS was shown to clean off railway track, within seconds, the organic layer produced when leaves are crushed onto it, whilst ultrasonic cleaning baths were unable to remove such a contaminant, even after many minutes of immersion. In addition to the advantages established above (UAS could clean the track in place, whilst the track had to be cut in into sections small enough to fit into the bath, transported to it, and were immersed in it off line), UAS brought greater effectiveness in cleaning whilst also being (in tests to date) 'hands safe'. This particular contamination is responsible for the 'leaves on the line' problem which costs the UK rail network £50M annually, as timetables are adjusted to reduce the speed of the trains to compensate for the decrease in braking ability that crushed leaf contamination causes.

Therefore comparison of these two ultrasonic cleaning methods provides fruitful discussions on effectiveness, whilst highlighting that there are particular scenarios to which one or other is suited (whilst the need to immerse targets brings many limitations to ultrasonic cleaning baths as discussed above, it does allow batch cleaning, and so can clean large numbers of small items more quickly than could a single UAS nozzle).

In contrast, there is little to be gained in comparing cleaning effectiveness of any system against that of the pressure washer, since the latter can always be scaled up, to the point where the most powerful will cut through thick steel. The issue therefore is not one of how well each cleans, but whether the scenario will allow deployment. Can the user cope with the slow rate of covering an area that a single UAS nozzle requires? Can the user accept the surface damage, aerosol production, and use of electrical and water resources that come with the size of pressure washer that is required to achieve an acceptable level of cleaning? Does the need to 'clean-in-place' preclude use of an ultrasonic bath? Section 4.3 will explore a particular application, the cleaning of wounds and ulcers, where a UAS nozzle (which would generate water hammer pressures 100,000 less than those generated by the weakest domestic pressure washer) might make an interesting comparison with pulsed lavage system [101], particularly if low water and power requirements were to make them portable.

4.3. Implications for the cleaning of wounds and ulcers

As mentioned above, StarStream will enhance the efficacy of additives (e.g. detergents or biocides) in the penetration of cracks and crevices, so why demonstrate (in figures 7-9) its cleaning ability with just cold water and no additives? Some targets (e.g. lettuce [102]) can only tolerate cold water. In other circumstances (e.g. rural wound cleaning in a Low/Middle Income Country [LMIC]) cold bottled or packaged water might be the only clean water available [103].

Even in controlled hospital conditions in the developed world, wound cleaning is a major and unsolved issue. Considering just diabetics (which 'is now the biggest cause of amputation' [104]), the inability to treat ulcers properly creates a huge cost to patients and to the NHS. Discussing the care of diabetics in England, Kerr [105] states that: "Around 6,000 people with diabetes undergo leg, foot or toe amputation each year in England. Many of these amputations are avoidable.... Around 61,000 people with diabetes are thought to have foot ulcers at any given time, approximately 2.5% of the diabetes population... Only 50% of patients with diabetes who have had an amputation survive for a further 2 years. Even without amputation, the prognosis is poor. Only around 56% of people with diabetes who have had ulcers survive for 5 years... In 2010-11, the NHS in England spent an estimated £639 million–£662 million, 0.6–0.7% of its budget, on diabetic foot ulceration and amputation." In particular, by calculating the excess bed days as the difference between the number of days that a procedure actually took and the number of days it would normally be expected to take, Kerr estimated that 'expenditure on excess bed days in people with diabetic foot ulcers admitted to non-foot care HRGs is estimated at £100 million... The annual cost of non-amputation inpatient care for diabetic foot ulcers is estimated at £213 million' (an HRG - Healthcare Resource Group - is grouping of patient events that consume a specified resource).

The impact of enhanced wound cleaning in the NHS that StarStream (or some other procedure) might provide could therefore be significant. However StarStream's emphasis on the use of low volumes of cold water has other implications, notably because it turns cold drinking water (which on its own is not highly effective as a wound cleaner) into a more effective cleaner. If this enhanced cleaning ability translated through to wound cleaning, StarStream would have particularly applicability for those scenarios in which clean water is a scarce, because water is a dense commodity that has multiple uses. An army medic, or mountain rescue or catastrophe zone worker, might carry a limited amount of drinking water, but appreciate having the ability (when the need arises) to pass his/her remaining drinking water reserves through a battery-powered StarStream nozzle (perhaps with powder added first to make saline) in order to clean a wound for a few minutes. Battlefield and catastrophe zone wounds can be extremely hazardous, and sepsis can set before the victim arrives at hospital. Cleaning the wound first can reduce that hazard. Ambulances, rescue vehicles, small boats, helicopters etc. could all carry such items. Of course, for much of the world the prospect of transportation to a modern hospital is remote even for as predictable an event as rural childbirth in peace time [106], and in much of the world healthcare needs are made less predictable and manageable by accident, migration, conflict, resource shortages and infrastructure that is poor or overwhelmed.

Sepsis is indeed a huge problem: Severe sepsis hospitalizations are currently doubling each decade, resulting in at least 750,000 persons affected annually in the United States [107, 108], and 500,000 patients with severe sepsis are treated annually in US emergency departments [109], 100,000 of which are children [110, 111]. In one study, 27.1% of a set of 56,673 adult admissions to Intensive Care in England, Wales and Northern Ireland between 1995 and 2000 met severe sepsis criteria within the first 24 hours [112]. These admissions accounted for 45% of the total usage of the intensive care unit, and 33% of the hospital bed days used by all intensive care unit admissions. Of these, 35% died before being discharged from the intensive care unit, and the proportion that died at some point during their hospital stay was 47% [112]. Another study evaluated 28,150 patients with severe sepsis and septic shock, from January 2005 through February 2010, and found that in-hospital mortality was 29.7% for the cohort as a whole [113]. This percentage includes 17,990 patients who received antibiotics after sepsis identification. Expert consensus suggested that every 1 hour delay in effective therapy decreases survival by around 6%, and the Surviving Sepsis Guidelines mandate that intravenous antimicrobial therapy must be administered within 1 hour of recognizing septic shock [113, 114-119].

What has this to do with cold water cleaning? The above Surviving Sepsis Guidelines are challenging to implement [120]. Levy *et al.* [121] analysed data from 165 hospitals treating over 15,000 patients with septic shock and revealed that only 68% of patients received broad-spectrum antibiotics within 3 hours of presentation at the Emergency Department. However, sepsis is often not caused by infection (and so will not respond to antimicrobial therapy). Even if it is caused by infection, those infections may be fungal or viral, and not respond to antibiotics. Therefore the pressure outlined above to apply a broad-spectrum antibiotic may produce possible unwanted effects (such as suppression of the gut microbiome), could displace the required therapy that would be used had the cause of sepsis been properly identified, and will promote the development of Antibiotic Resistance. Cleaning wounds on-site, before transport to hospital, might therefore have wider benefits than simply reducing the hazard of sepsis in the patient – on a large scale it is a tactic to mitigate against the development of AntiMicrobial resistance (AMR), which is an arms race against natural selection that cannot be won [50]. To mitigate against potential catastrophes in healthcare and food production (which currently relies upon antibiotics to farm sufficient livestock to feed large tracts of the world), measures over and above the development of new antibiotics must be undertaken [50]. Two key elements of this are:

- infection prevention - if the microbe never enters the body, no antimicrobial is required;
- the removal of environments that encourage resistant strains to develop: the longer microbes are allowed to persist on surfaces involved in food preparation, infection management and water and waste treatment, the increased their chance of developing resistance and/or transferring it between microbes.

Cleaning is key to both these elements, and applies not just to hospital infections. To take just one example, despite all efforts, the UK population persists in washing their hands lamentably short of the recommended 20 s in warm soapy water. The past decade, and more, has shown that we cannot change the behaviour of sufficient numbers of people with campaigns to educate them on hand washing: the StarStream philosophy would be that, if we cannot change the behaviour, we change the water, making any shorter washes in cold water as effective as they can be.

StarStream cleans by a mechanical scrubbing action, which is very different to the chemical attacks that microbes are often challenged with, so giving another route to combat AMR. A major problem is that such agents enter the environment (the water supply, sewage, run-off etc.) where their occurrence aids the development of AMR. Microbe populations can become resistant because traditional chemical routes to cleaning leave 'smoking guns' in our water run-off, giving clues as to what was used to kill the previous generations of microbes. These 'clues' in sub-therapeutic concentrations might assist the vast reservoirs of microbes in our water and sewage systems to develop resistance. With StarStream, not only can we clean well, but nothing enters the run-off except the material we rinse off – there are no chemical clues that are flushed into the environment that microbe populations can use in the development of AMR.

The implications go further. The scenario raised above, of rescue personnel and vehicles needing to conserve the volume of water they carry but undertake adequate wound cleaning, will be even more pertinent in a decade, when the first responder in many incidents is likely to be a drone. With its powerful battery but limitations in the amount of liquid it can carry, a drone makes an ideal test scenario for UAS: whilst it would be logical for a drone to carry a small amount of water to rehydrate a person needing assistance, a trade-off would normally need to be made between carrying liquid for rehydration and carrying liquid for wound cleaning. The ability to undertake cold water cleaning with water that could otherwise be used for rehydration ameliorates the conflict inherent in the choice of which liquid to carry. Indeed, by making drinking water dual-use (for rehydration and wound cleaning), there is impetus to build not-for-profit battery-powered reusable UAS nozzles that can fit on the top of any bottle of drinking water, for those parts of the world (in rural clinics, maternity units etc.) where bottled water represents the only source of clean water: UAS would turn such water from being a mediocre wound cleaner into a greatly enhanced one (Figure 11).

(a) (b) (c)

Figure 11. Concept for a battery-powered solar-charged UAS device that can fit onto any bottled drinking water unit. For the video see https://youtu.be/o903Yey71L4

6. Conclusions

The interaction of bubbles with acoustic fields provides a route by which fundamental research, covering physical, engineering, zoology, biophysics and biomedicine can be conducted and translated through to benefit society.

A key opportunity for bubble acoustics is in cold water cleaning, and its potential to address AMR. Unless preventative measures are found (and no-one in the world currently knows what those will be), AMR will (through the colloquial 'rise of superbugs') by 2050 be killing more people than cancer, and cost the world economy more than the current size of the global economy. We will not be able to feed the world unless we wean our food production industry off its dependence on antibiotics; common medical procedures (minor surgery, childbirth) will become significantly more hazardous; and advances in treatments (such as those for childhood leukaemia) will become reversed.

To offset this doom, the benefits for finding these preventative measures now would revolutionise current healthcare, the example given here being of the revolution to rural childbirth, wound treatment and childbirth, and the cost of amputations in diabetics.

Acknowledgements

The literature search for AMR contained in section 4.3 was funded by EPSRC's Network for Antimicrobial Action 'Bridging the Gap' grant NAMRA (EP/M027260/1), and by the University of

Southampton Strategic Research Group NAMRIP http://www.southampton.ac.uk/namrip/index.page?. The author is very grateful to Craig Dolder and Mengyang Zhu for assistance in generating figure 10, and to Dr Dolder for drawing figure 11 at the author's request, and for animating it: the animation, and other data from study are openly available from the University of Southampton repository at is http://dx.doi.org/10.5258/SOTON/400818.

References

[1] Rayleigh, L 1917 On the pressure developed in a liquid during collapse of a spherical cavity *Phil. Mag*. **34**, 94-98

[2] Leighton T G 2004 From seas to surgeries, from babbling brooks to baby scans: The acoustics of gas bubbles in liquids *International Journal of Modern Physics B* **18**(25) 3267-314

[3] Noltingk B E and Neppiras E A. 1950 Cavitation produced by ultrasonics. *Proc. Phys. Soc.* B **63**, 674-685.

[4] Neppiras E A and Noltingk B E. 1951 Cavitation produced by ultrasonics: Theoretical conditions for the onset of cavitation. *Proc. Phys. Soc.* B **64**, 1032-1038.

[5] Minnaert M 1933 On musical air-bubbles and sounds of running water *Phil. Mag.* **16** 235-48

[6] Leighton T G and Walton A J 1987 An experimental study of the sound emitted from gas bubbles in a liquid *European Journal of Physics* **8** 98-104

[7] Pumphrey H C and Walton A J 1988 Experimental study of the sound emitted by water drops impacting on a water surface *Eur. J. Phys*. **9**, 225-31

[8] Leighton T G, White P R and Schneider M F 1998 The detection and dimension of bubble entrainment and comminution *J. Acoust. Soc. Am.* **103** 1825-35

[9] Nystuen J A and McPhaden M J 2001 The beginnings of operational marine weather observations using underwater ambient sound, *In Acoustical Oceanography,* ed TG Leighton, GJ Heald, H Griffiths and G Griffiths (Institute of Acoustics) *Proceedings of the Institute of Acoustics* **23**(2) 135-41

[10] Medwin H and Daniel A C Jr. 1990 Acoustical measurements on bubble production by spilling breakers *J. Acoust. Soc. Am.* **88** 408-12

[11] Updegraff G E and Anderson V C 1991 Bubble noise and wavelet spills recorded 1 m below the ocean surface *J. Acoust. Soc. Am.* **89** 2264-79

[12] Thorpe S 1982 On the clouds of bubbles formed by breaking wind-waves in deep water, and their role in air-sea gas transfer *Philos. Trans. Roy. Soc. Lond. A* **304** 155-210

[13] Farmer D M and Vagle S 1989 Waveguide propagation of ambient sound in the ocean-surface bubble layer *J. Acoust. Soc. Am.* **86** 1897-1908

[14] Leighton T G, Lingard R J, Walton A J and Field J E 1991 Acoustic bubble sizing by the combination of subharmonic emissions with an imaging frequency *Ultrasonics* **29**, 319-23

[15] Leighton TG, Phelps A D, Ramble D G and Sharpe D A 1996 Comparison of the abilities of eight acoustic techniques to detect and size a single bubble *Ultrasonics* **34** 661-7

[16] Leighton T G, Ramble D G and Phelps A D 1997 The detection of tethered and rising bubbles using multiple acoustic techniques *J. Acoust. Soc. Am.* **101**(5) 2626-35

[17] Phelps A D and Leighton T G 1997 The subharmonic oscillations and combination-frequency emissions from a resonant bubble: their properties and generation mechanisms. *Acta Acustica* **83** 59-66

[18] Phelps A D and Leighton T G 1996 High resolution bubble sizing through detection of the subharmonic response with a two frequency excitation technique', J. Acoust. Soc. Am. 99, 1985-92. (1996).

[19] Ramble D G, Phelps A D and Leighton T G 1998 On the relation between surface waves on a bubble and the subharmonic combination-frequency emission *Acustica with Acta Acustica* **84**(5), 986-88

[20] Phelps A D, Ramble D G and Leighton T G 1997 The use of a combination frequency technique to measure the surf zone bubble population *J. Acoust. Soc.Am.* **101**(4) 1981-9

[21] Phelps A D and Leighton TG 1998 Oceanic bubble population measurements using a buoy-deployed combination frequency technique *IEEE Journal of Oceanic Engineering* **23**(4), 400-10

[22] Leighton T G, Meers S D and White P R 2004 Propagation through nonlinear time-dependent bubble clouds and the estimation of bubble populations from measured acoustic characteristics *Proc. Roy. Soc. Lond. A* **460**(2049) 2521-50

[23] Clarke J W L and Leighton T G 2000 A method for estimating time-dependent acoustic cross-sections of bubbles and bubble clouds prior to the steady state *J. Acoust. Soc. Am.* **107**(4), 1922-9

[24] Leighton T G and White P R 2014 Dolphin-inspired target detection for sonar and radar *Archives of Acoustics*, **39**(3) 319-32 doi: 10.2478/aoa-2014-0037

[25] Leighton T G, Finfer D C, White P R, Chua G-H and Dix J K 2010 Clutter suppression and classification using Twin Inverted Pulse Sonar (TWIPS) *Proceedings of the Royal Society A*, **466**, 3453-78 doi:10.1098/rspa.2010.0154

[26] Leighton, T G, Finfer D C, Chua G H, White P R and Dix J K 2011 Clutter suppression and classification using Twin Inverted Pulse Sonar in ship wakes *Journal of the Acoustical Society of America* **130**(5) 3431-7 doi:10.1121/1.3626131

[27] Leighton T G, Chua G-H and White P R 2012 Do dolphins benefit from nonlinear mathematics when processing their sonar returns? *Proceedings of the Royal Society A* **468** 3517-32 doi: 10.1098/rspa.2012.0247

[28] Chua, G-H, White P R and Leighton T G 2012 Use of clicks resembling those of the Atlantic bottlenose dolphin (*Tursiops truncatus*) to improve target discrimination in bubbly water with biased pulse summation sonar *IET Radar Sonar & Navigation* **6**(6) 510-15 doi: 10.1049/iet-rsn.2011.0199

[29] Finfer D C, White P R, Chua G H and Leighton T G 2012 Review of the occurrence of multiple pulse echolocation clicks in recordings from small odontocetes *IET Radar Sonar & Navigation* **6**(6) 545-55 doi: 10.1049/iet-rsn.2011.0348

[30] Leighton T G, Chua G-H, White P R, Tong K F, Griffiths H D and Daniels D J 2013 Radar clutter suppression and target discrimination using twin inverted pulses *Proceedings of the Royal Society London A* **469**(2160) 20130512 (14 pages) doi: 10.1098/rspa.2013.0512

[31] Leighton T G, Richards S D and White P R 2004 Trapped within a 'wall of sound': A possible mechanism for the bubble nets of the humpback whales *Acoustics Bulletin*, **29**(1), 24-29

[32] Leighton T G, Finfer D, Grover E and White P R 2007 An acoustical hypothesis for the spiral bubble nets of humpback whales and the implications for whale feeding *Acoustics Bulletin*, **22**(1), 17-21

[33] Leighton T G, Finfer D and White P R 2007 Cavitation and cetacean *Revista de Acustica*, **38**(3/4), 37-81

[34] Leighton T G and White P R 2004 The Sound of Titan: A role for acoustics in space exploration *Acoustics Bulletin* **29**(4) 16-23

[35] Leighton T G and Petculescu A 2008 Sounds in space: The potential uses for acoustics in the exploration of other worlds *Hydroacoustics* **11** 225-38

[36] Jiang J, Baik K and Leighton T G 2011 Acoustic attenuation, phase and group velocities in liquid-filled pipes II: Simulation for spallation neutron sources and planetary exploration *J. Acoust. Soc. Am.* **130**(2) 695-706

[37] Leighton T G 2009 Fluid loading effects for acoustical sensors in the atmospheres of Mars, Venus, Titan and Jupiter *J. Acoust. Soc. Am.* **125**(5), EL214-EL219

[38] Ainslie M A and Leighton T G 2016 Sonar equations for planetary exploration. *Journal of the Acoustical Society of America*, **140**(2), 1469–1480 (doi:10.1121/1.4960785).

[39] Ainslie M A and Leighton T G 2009 Near resonant bubble acoustic cross-section corrections, including examples from oceanography, volcanology, and biomedical ultrasound *Journal of the Acoustical Society of America*, **126**(5), 2163-2175

[40] Robb G B N, Robinson S P, Theobald P D, Hayman G, Humphrey V F, Leighton T G, Wang L S, Dix J K and Best A I 2009 Absolute calibration of hydrophones immersed in sandy sediment *Journal of the Acoustical Society of America*,**125**(5), 2918-2927

[41] Leighton T G, Finfer D C and White P R 2008 The problems with acoustics on a small planet *Icarus* **193**(2) 649-52

[42] Leighton T G, White P R and Finfer D C 2012 The opportunities and challenges in the use of extra-terrestrial acoustics in the exploration of the oceans of icy planetary bodies *Earth Moon and Planets* **109**(1-4), 91-116

[43] Leighton T G 2012 The use of extra-terrestrial oceans to test ocean acoustics students', J. Acoust. Soc. Am. 131(3 Pt 2), 2551-5.

[44] Petculescu A and Leighton T G 2016 Guest editorial: Acoustic and related waves in extraterrestrial environments. *Journal of the Acoustical Society of America*, **140**(2), 1397–1399 (doi:10.1121/1.4961539).

[45] Leighton T G, Banda N, Berges B, Joseph PF and White P R 2016 Extraterrestrial sound for planetaria: a pedagogical study. *Journal of the Acoustical Society of America*, **140**(2), 1400-1419 (doi:10.1121/1.4960786)

[46] Leighton T G and Petculescu A 2009 The sound of music and voices in space Part 1: Theory *Acoustics Today* **5**(3) 17-26

[47] Leighton T G and Petculescu A 2009 The sound of music and voices in space Part 2: Modelling and simulation *Acoustics Today* **5**(3) 27-9

[48] Leighton T G 2015 The sounds of voices and waterfalls on other planets *University of Southampton* http://tinyurl.com/znx4rve (publ. Dec 13 2014; Last accessed Feb 21, 2016)

[49] Cook J.-R., Brown D C and Laustsen P 2010 Cassini Spots Potential Ice Volcano on Saturn Moon," Cassini Solstice Mission, News and Features (Dec 14, 2010), Jet Propulsion Lab California Inst. Tech. http://saturn.jpl.nasa.gov/news/newsreleases/newsrelease20101214/ (last observed 5 Dec 2015).

[50] Leighton T G 2015 We need more than just new antibiotics to fight superbugs, The Conversation (15 July 2015) https://theconversation.com/we-need-more-than-just-new-antibiotics-to-fight-superbugs-44054 (publ. March 5, 2015; Last accessed Feb 21, 2016)

[51] Leighton T G, Birkin P R and Offin D. 2013 A new approach to ultrasonic cleaning, *Proceedings of the International Congress on Acoustics*, **19**, paper 075029 (2013) (4 pages) doi: 10.1121/1.4799209

[52] Enk M D and Mathis B J 1977 Distribution of cadmium and lead in a stream ecosystem *Hydrobiologia* **52**(2-3) 153-8

[53] Clark E H 1985 The off-site costs of soil erosion *J. Soil and Water Conservation* **40**(1) 19-22

[54] Cohen E 2015 Deadly superbug-related scopes sold without FDA approval. CNN News Website http://edition.cnn.com/2015/03/04/us/superbug-endoscope-no-permission/ (published March 5, 2015; Last accessed Feb 21, 2016)

[55] Shaw M C 2015 Who's Watching The Watchers: Big Problems With ERCP Scopes. Health News Digest http://www.healthnewsdigest.com/news/contributing%20columnist0/Who-s-Watching-The-Watchers-Big-Problems-With-ERCP-Scopes.shtml (published Sep 13, 2015; Last accessed Feb 21, 2016)

[56] Leighton T G 2015 We need more than just new antibiotics to fight superbugs, *The Conversation* https://theconversation.com/we-need-more-than-just-new-antibiotics-to-fight-superbugs-44054 (published 15 July 2015; Last accessed Feb 21, 2016)
[57] Leighton T G 2016 Are some people suffering as a result of increasing mass exposure of the public to ultrasound in air? *Proc. Roy. Soc. A* **472**(2185) 20150624 (57 pages)
[58] Maksimov A O and Leighton T G 2001 Transient processes near the threshold of acoustically driven bubble shape oscillations *Acta Acustica* **87**(3) 322-332
[59] Maksimov A O, Leighton T G and Birkin P R 2008 Self focusing of acoustically excited Faraday ripples on a bubble wall *Physics Letters A* **372**(18) 3210-16
[60] Birkin P R, Offin D G, Vian C J B, Leighton T G and Maksimov AO 2011 Investigation of non-inertial cavitation produced by an ultrasonic horn *J. Acoustical Society of America* **130**(5) 3297-3308 doi:10.1121/1.3650537
[61] Maksimov A O and Leighton T G 2012 Pattern formation on the surface of a bubble driven by an acoustic field *Proc. Roy. Soc. A* **468** 57-75 (doi:10.1098/rspa.2011.0366)
[62] Leighton T G, Ho W L and Flaxman R 1997 Sonoluminescence from the unstable collapse of a conical bubble *Ultrasonics* **35** 399-405
[63] Leighton T G, Phelps A D, Cox B T and Ho W L 1998 Theory and preliminary measurements of the Rayleigh-like collapse of a conical bubble *Acustica with Acta Acustica* **84**(6) 1014-24
[64] Leighton T G, Cox B T and Phelps A D 2000 The Rayleigh-like collapse of a conical bubble *J. Acoustical Society of America* **107**(1) 130-142
[65] Birkin P R, Power J F and Leighton T G 2001 Electrochemical evidence of H produced by ultrasound *J. Chem. Soc. Chemical Communication*, **21** 2230-1
[66] Birkin P R, Power J F, Leighton T G and Vincotte, AML 2002 Cathodic electrochemical detection of sonochemical radical products *Analytical Chemistry* **74** 2584-90
[67] Birkin P R, Offin D G, Joseph P F and Leighton T G 2005 Cavitation, shock waves and the invasive nature of sonoelectrochemistry *J. Physical Chemistry B* **109** 16997-17005
[68] Birkin P R, Offin D G, and Leighton T G 2005 Experimental and theoretical characterisation of sonochemical cells. Part 2: Cell disruptors (ultrasonic horn) and cavity cluster collapse *Physical Chemistry Chemical Physics* , 530-37
[69] Turangan C K, Jamaluddin A R, Ball G J and Leighton T G 2008 Free-Lagrange simulations of the expansion and jetting collapse of air bubbles in water *Journal of Fluid Mechanics* **598** 1-25 doi:10.1017/S0022112007009317
[70] Jamaluddin A R, Ball G J, Turangan C K and Leighton T G 2011 The collapse of single bubbles and calculations of the far-field acoustic emissions for cavitation induced by shock wave lithotripsy *Journal of Fluid Mechanics* **677** 305-41 doi:10.1017/jfm.2011.85
[71] Birkin P R, Offin D G, and Leighton T G 2004 Electrochemical measurements of the effects of inertial acoustic cavitation by means of a novel dual microelectrode *Electrochemistry Communications* **6**(11) 1174-79
[72] Leighton T G, Turangan C K, Jamaluddin A R, Ball G J and White P R 2013 Prediction of far-field acoustic emissions from cavitation clouds during shock wave lithotripsy for development of a clinical device *Proceedings of the Royal Society* **469**(2150) 20120538, 21pp (doi: 10.1098/rspa.2012.0538).
[73] Pickworth M J W, Dendy P P, Twentyman P R and Leighton T G 1989 Studies of the cavitational effects of clinical ultrasound by sonoluminescence: 4. The effect of therapeutic ultrasound on cells in monolayer culture in a standing wave field *Physics in Medicine and Biology* **34**(11) 1553-60
[74] Birkin P R, Leighton T G, Power J F, Simpson M D, Vincotte A M L and Joseph P 2003 Experimental and theoretical characterisation of sonochemical cells. Part 1. Cylindrical reactors and their use to calculate the speed of sound in aqueous solutions *J. Physical Chemistry A* **107** 306-20

[75] Faraday M 1831 On the forms and states assumed by fluids in contact with vibrating elastic surfaces *Phil. Trans. Roy. Soc. London* **121** 319-40
[76] Watson Y E, Birkin P R and Leighton T G 2003 Electrochemical detection of bubble oscillation *Ultrasonics Sonochemistry* **10** 65-9
[77] Elder S A 1958 Cavitation microstreaming *J. Acoust. Soc. Am.* **31** 54-64.
[78] Nyborg W L 1958 Acoustic streaming near a boundary *J. Acoust. Soc. Am.* **30** 329-39
[79] Birkin P R, Watson Y E, Leighton T G and Smith K L 2002 Electrochemical detection of Faraday waves on the surface of a gas bubble *Langmuir Surfaces and Colloids* **18** 2135-40
[80] Rooney J A 1970 Hemolysis near an ultrasonically pulsating gas bubble *Science* **169** 869
[81] Lewin P A and Bjørnø L 1982 Acoustically induced shear stresses in the vicinity of microbubbles in tissue *J. Acoust. Soc. Am.* **71** 728
[82] Offin D G, Birkin P R and Leighton T G 2007 Electrodeposition of copper in the presence of an acoustically excited gas bubble *Electrochemistry Communications* **9**(5) 1062-68
[83] Leighton T G, Birkin P R and Vian C 2016 Cleaning Apparatus and Method, and Monitoring Thereof. European Patent Number EP 2470310 from Application No. 10748081.6 Ultrasonic wave device. University Of Southampton (granted on 06/01/2016; filed on 24 February 2012 with an allocated filing date on 26 August 2010) Published 04 July 2012 under 2470310. Claiming priority from GB 0914836.2 and PCT/EP2010/062448. From UK Patent Application No. 0914836.2 (University of Southampton 2009, filed 26 August 2009, published 26 August 2010).
[84] Leighton T G 1994 *The Acoustic Bubble,* Academic Press, 640pages (ISBN 0124419208).
[85] Birkin P R, Offin D G, Vian C J B, Howlin R P, Dawson J I, Secker T J, Herve R C, Stoodley P, Oreffo R O C, Keevil C W and Leighton T G 2015 Cold water cleaning of brain proteins, biofilm and bone - harnessing an ultrasonically activated stream *Physical Chemistry Chemical Physics* **17** 20574-79 doi: 10.1039/C5CP02406D
[86] Howlin R P, Fabbri S., Offin D G, Symonds N, Kiang K S, Knee R J, Yoganantham D C, Webb J S, Birkin P R, Leighton T G and Stoodley P 2015 Removal of dental biofilms with a novel ultrasonically-activated water stream *J. Dental Research* **94**(9) 1303-09 doi:10.1177/0022034515589284
[87] Birkin P R, Offin D G, Vian C J B and Leighton T G 2015 Electrochemical "bubble swarm" enhancement of ultrasonic surface cleaning *Physical Chemistry Chemical Physics* **17**(33) 21709-21715 (doi:10.1039/c5cp02933c). (PMID:26234563).
[88] Salta M, Goodes L, Mass B, Dennington S, Secker T, Leighton TG 2016 Bubbles vs. biofilms: A novel method for the removal of marine biofilms attached on antifouling coatings using an ultrasonically activated water stream. *Surface Topography: Metrology and Properties,* **4**(3), 034009 doi: 10.1088/2051-672X/4/3/034009.
[89] Goodes, L., Harvey, T., Symonds, N. and Leighton, T.G. 2016 A comparison of ultrasonically activated water stream and ultrasonic bath immersion cleaning of railhead leaf-film contaminant. *Surface Topography: Metrology and Properties*, **4**(3), 034004 doi: 10.1088/2051-672X/4/3/034003.
[90] Leighton T G 2015 The acoustic bubble: Oceanic bubble acoustics and ultrasonic cleaning. *Proceedings of Meetings on Acoustics (POMA), Acoustical Society of America,* **24**(1) doi: 10.1121/2.0000121
[91] Agarwal A, Ng W J and Liu Y 2011 Principle and applications of microbubble and nanobubble technology for water treatment *Chemosphere* **84**, 1175–1180
[92] Wu Z, Chen H, Dong Y, Mao H, Sun J, Chen S, Craig V S J and Hu J 2008 Cleaning using nanobubbles: Defouling by electrochemical generation of bubbles *J. Colloid and Interface Science* **328** (2008) 10–14
[93] Yang S and Duisterwinkel A 2011 Removal of nanoparticles from plain and patterned surfaces using nanobubbles *Langmuir*, **27** (18), 11430–11435

[94] Liu G and Craig V S J 2009 Improved cleaning of hydrophilic protein-coated surfaces using the combination of nanobubbles and SDS *ACS Appl. Mater. Interfaces*, **1**(2), 481-487

[95] Leighton, T.G., Walton, A.J. and Pickworth, M.J.W 1990 Primary Bjerknes forces, *European Journal of Physics*, **11**, 47-50

[96] Leighton T G 1994 *The Acoustic Bubble,* Academic Press, 640 pages (ISBN 0124419208), pp. 341-367.

[97] Offin, D.G., Birkin, P.R. and Leighton, T.G. (2014) An electrochemical and high-speed imaging study of micropore decontamination by acoustic bubble entrapment, *Phys. Chem. Chem. Phys.*,**16**(10), 4982-4989 doi:10.1039/C3CP55088E.

[98] Leighton T G, Birkin P R, Hodnett M, Zeqiri B, Power J F, Price G J, Mason T, Plattes M, Dezhkunov N and Coleman A J 2005 Characterisation of measures of reference acoustic cavitation (COMORAC): An experimental feasibility trial. In *Bubble and Particle Dynamics in Acoustic Fields: Modern Trends and Applications* ed A.A. Doinikov (Research Signpost, Kerala, Research Signpost) 37-94

[99] Leighton T G 2007 What is ultrasound? *Progress Biophysics and Molecular Biol.* **93**(1-3) 3-83.

[100] Goodes, L., Harvey, T., Symonds, N. and Leighton, T.G. 2016 The Adhesion Rail Riddle – Ensuring trains can brake. University of Southampton (last accessed 23 Sept 2016) http://www.southampton.ac.uk/engineering/research/projects/the-adhesion-rail-riddle.page

[101] Luedtke-Hoffmann K A and Schafer D S 2000 Pulsed lavage in wound cleansing. *Physical Therapy*, **80**(3), 292–300.

[102] Leighton, T.G. 2016 StarStream (2007-2015) - Cleaning with low volumes of cold water. University of Southampton (last accessed 23 Sept 2016) http://www.southampton.ac.uk/engineering/research/projects/starstream.page?

[103] Safe Water Network 2016 Ghana (last accessed 23 Sept 2016) http://www.safewaternetwork.org/countries-regions/ghana

[104] McInnes A D (2012) Diabetic foot disease in the United Kingdom: about time to put feet first. *J. Foot and Ankle Research*, **5**, 26-32.

[105] Kerr M, 2011 Insight Health Economics: Inpatient Care for People with Diabetes: The Economic Case for Change. UK: NHS Diabetes.

[106] Bantebya Kyomuhendo G 2003 Low use of rural maternity services in Uganda: Impact of women's status, traditional beliefs and limited resources. *Reproductive Health Matters,* **11**(21), 16–26.

[107] Angus D C, Linde-Zwirble W T, Lidicker J, Clermont G, Carcillo J, Pinsky M R 2001 Epidemiology of severe sepsis in the United States: analysis of incidence, outcome, and associated costs of care. *Critical Care Medicine* **29**(7) 1303–10 PubMed: 11445675

[108] Dombrovskiy V Y, Martin A A, Sunderram J and Paz H L 2007 Rapid increase in hospitalization and mortality rates for severe sepsis in the United States: A trend analysis from 1993 to 2003. *Critical Care Medicine* **35**(5) 1244–50 PubMed: 17414736

[109] Wang H E, Shapiro N I, Angus D C and Yearly D M 2007 National estimates of severe sepsis in United States emergency departments *Crit. Care. Med.* **35**(8) 1928–36.

[110] Singhal S, Allen M W, McAnnally J R, Smith K S, Donnelly J P and Wang H E 2013 National estimates of emergency department visits for pediatric severe sepsis in the United States *PeerJ.* **1** e79. doi: 10.7717/peerj.7

[111] Horeczko T, Green J P and Panacek E A 2014 Epidemiology of the Systemic Inflammatory Response Syndrome (SIRS) in the emergency department *West J Emerg Med.* **15**(3) 329-36 doi: 10.5811/westjem.2013.9.18064

[112] Padkin A, Goldfrad C, Brady A, Young D, Black N and Rowan K 2003 Epidemiology of severe sepsis occurring in the first 24 hrs in intensive care units in England, Wales, and Northern Ireland *Critical Care Medicine* **31**(9) 2332-38 doi: 10.1097/01.CCM.0000085141.75513.2B

[113] Ferrer R, Martin-Loeches I, Phillips G, Osborn T, Townsend S, Dellinger R P, Artigas A, Schorr C and Levy M 2014 Empiric antibiotic treatment reduces mortality in severe sepsis and septic shock from the first hour: results from a guideline-based performance improvement program *Crit. Care. Med.* **42**(9) 1749-55 doi: 10.1097/CCM.0000000000000330

[114] Gaieski D F, Mikkelsen M E, Band R A, Pines J M, Massone R, Furia F F, Shofer F S and Goyal M 2010 Impact of time to antibiotics on survival in patients with severe sepsis or septic shock in whom early goal-directed therapy was initiated in the emergency department *Critical Care Medicine* **38**(4) 1045-53 doi: 10.1097/CCM.0b013e3181cc4824

[115] Dellinger R P, Levy M M, Rhodes A, Annane D, Gerlach H, Opal S M, Sevransky J E, Sprung C L, Douglas I S, Jaeschke R, Osborn T M, Nunnally M E, Townsend S R, Reinhart K, Kleinpell R M, Angus D C, Deutschman C S, Machado F R, Rubenfeld G D, Webb S, Beale R J, Vincent J L, Moreno R and the Surviving Sepsis Campaign Guidelines Committee including The Pediatric Subgroup. 2013 Surviving Sepsis Campaign: international guidelines for management of severe sepsis and septic shock, 2012. *Intensive Care Med.* **39**(2) 165-228. doi: 10.1007/s00134-012-2769

[116] Dellinger R P, Levy M M, Rhodes A, Annane D, Gerlach H, Opal S M, Sevransky J E, Sprung C L, Douglas I S, Jaeschke R, Osborn T M, Nunnally M E, Townsend S R, Reinhart K, Kleinpell R M, Angus D C, Deutschman C S, Machado F R, Rubenfeld G D, Webb S A, Beale R J, Vincent J L, Moreno R and the Surviving Sepsis Campaign Guidelines Committee including the Pediatric Subgroup 2013 Surviving sepsis campaign: international guidelines for management of severe sepsis and septic shock: 2012. *Critical Care Medicine* **41**(2) 580-637 doi: 10.1097/CCM.0b013e31827e83af. PMID: 23353941

[117] Levy M M, Dellinger R P, Townsend S R, Linde-Zwirble W T, Marshall J C, Bion J, Schorr C, Artigas A, Ramsay G, Beale R, Parker M M, Gerlach H, Reinhart K, Silva E, Harvey M, Regan S and Angus D C 2010 The Surviving Sepsis Campaign: results of an international guideline-based performance improvement program targeting severe sepsis. *Intensive Care Med.* **36**(2) 222-31 doi: 10.1007/s00134-009-1738-3

[118] Levy M M, Dellinger R P, Townsend S R, Linde-Zwirble W T, Marshall J C, Bion J, Schorr C, Artigas A, Ramsay G, Beale R, Parker M M, Gerlach H, Reinhart K, Silva E, Harvey M, Regan S, Angus D C and Surviving Sepsis Campaign. 2010 The Surviving Sepsis Campaign: results of an international guideline-based performance improvement program targeting severe sepsis. *Crit. Care. Med.* **38**(2):367-74. doi: 10.1097/CCM.0b013e3181cb0cdc

[119] Kumar A, Roberts D, Wood K E, Light B, Parrillo J E, Sharma S, Suppes R, Feinstein D, Zanotti S, Taiberg L, Gurka D, Kumar A and Cheang M 2006 Duration of hypotension before initiation of effective antimicrobial therapy is the critical determinant of survival in human septic shock *Critical Care Medicine* **34**(6) 1589–96 PubMed: 16625125

[120] Puskarich M A, Trzeciak S, Shapiro NI, Arnold RC, Horton JM, Studnek JR, Kline JA and Jones AE 2011 Association between timing of antibiotic administration and mortality from septic shock in patients treated with a quantitative resuscitation protocol *Critical Care Medicine* **39**(9) 2066-71 doi:10.1097/CCM.0b013e31821e87ab

[121] Levy M M, Dellinger R P, Townsend S R, Linde-Zwirble W T, Marshall J C, Bion J, Schorr C, Artigas A, Ramsay G, Beale R, Parker M M, Gerlach H, Reinhart K, Silva E, Harvey M, Regan S, Angus D C and Surviving Sepsis Campaign 2010 The Surviving Sepsis Campaign: results of an international guideline-based performance improvement program targeting severe sepsis *Critical Care Medicine* **38**(2) 367–74 PubMed: 20035219

IOP Conf. Series: Journal of Physics: Conf. Series **797** (2017) 012002 doi:10.1088/1742-6596/797/1/012002

Geometric distortion of area in medical ultrasound images

T Bland[1], J Tong[2], B Ward[2] and N G Parker[1]

[1] School of Mathematics and Statistics, Newcastle University, Newcastle upon Tyne, UK
[2] Regional Medical Physics Department, Freeman Hospital, Newcastle upon Tyne, UK

E-mail: nick.parker@ncl.ac.uk

Abstract. Medical ultrasound scanners are typically calibrated to a speed of sound corresponding to the soft tissue average of 1540 m s^{-1}. In regions of different sound speed, for example, organs and tumours, the B-mode image becomes geometrically distorted from the true tissue cross-section, due to refraction and the misrepresentation of length. A ray model is developed to predict this distortion for a generalized two-dimensional object with atypical speed of sound, and verified against ultrasound images of a test object. We quantify the areal image distortion as a function of the key dependencies, including the speed of sound mismatch, the scanning format, the object size and its elongation. Our findings show that the distortion of area can be significant, even for relatively small speed of sound mismatches. For example, a 5% speed mismatch typically leads to a 10 – 20% distortion in area. These findings have implications for the accuracy of ultrasound-based evaluation of area and volume.

1. Introduction

Medical ultrasound scanners conventionally assume a uniform speed of sound throughout the body, corresponding to the soft tissue average of 1540 m s^{-1} [1]. However, bodily soft tissue is acoustically inhomogeneous and its speed of sound can vary by several percent across different regions [2], e.g., typical speeds of sound in fat, muscle and liver are 1440, 1580 and 1590 m s^{-1}, respectively [3]. These variations in speed of sound are detrimental to the image formation and the accuracy of its representation of the true tissue profile. For example, they induce aberrations in the acoustic wavefront, which degrades the formation and steering of the ultrasonic beam, and in turn reduces the spatial and contrast resolution of the image [4, 5, 6]; the correction of these aberrations is an active area of research (see, for example, [7, 8, 9, 10, 11]). Another implication of such speed variations is that the scanner overlooks refraction of the beam and misregisters the beam path length; this introduces geometric distortions into the image whereby the position of certain echoes will deviate from the actual locations of the features which generated them. While the presence of these geometric distortions is well known within the ultrasound community [12, 13, 14, 15, 16, 18, 19], they are usually deemed negligible for the purposes of qualitative diagnostics. However, there is increasing use of ultrasound imaging for quantitative measurements, notably the rapid developments in ultrasound-based evaluation of the volume of organs and tumours [20, 21, 22]. These geometric distortions introduce systematic errors in such measurements and are a key limiting factor in their diagnostic accuracy [2, 23], and yet this distortion remains largely unexplored as a source of error in area and volume measurements.

To illustrate this effect, consider an ambient medium with speed of sound v_1 and an ultrasound scanner which assumes the same value. An object with atypical speed of sound $v_2 \neq v_1$ is embedded within the medium and a B-mode image taken of its cross-section. Then, the distance within the object will be misrepresented in the image and the refraction at the object boundary will not be accounted for, both of which will contribute to the geometric distortion of the object's image. While this distortion has been predicted previously [13, 14, 15, 16, 18, 17, 19], a quantitative understanding remains lacking. This work aims to address this issue by quantifying the distortion of area of generalized objects in two-dimensional B-mode images. This is a first step in understanding the distortion of volume in 3D ultrasound images, which are typically reconstructed from 2D images [33, 34, 35]; then, if the object in question has atypical speed of sound, the above 2D distortion will introduce errors in the calculated area of each scanning plane, and thus the total calculated volume.

A ray model (described in Section 2) forms the basis of our work. Despite its simplicity, it is well-suited to reveal the underlying mapping of the object boundary to its image, the main focus of this work. Our work may be considered as an extension to that of Ref. [36], which employed a similar methodology to model how points (rather than objects) become geometrically distorted due to an atypical ambient speed of sound. We develop our model across the most common medical transducer arrays: linear, curved, phased, and vector arrays (following the categorization of Ref. [36]). Comparison to ultrasound images of phantoms (Section 3) confirm the success of the ray model in predicting the shape of the image. We quantify the distortion as a function of the speed of sound ratio v_2/v_1, the object's shape and size, and the array type. Our results show that a small speed of sound mismatch can lead to significant distortions: for example, a 5% mismatch in the speeds of sound can lead to around 15% error in the area of a circular object, and this is sensitive to the size and shape of the object. We conclude by discussing the wider implications of our findings within the context of diagnostic evaluation of area and volume.

2. Theory

2.1. Ray Model

We consider an elliptical (two-dimensional) object in the x-z plane, with uniform speed of sound v_2, embedded in an ambient medium with uniform speed of sound v_1. The object has semi-widths b and c. The set-up is depicted in figure 1(a). The scanner is calibrated to a speed of sound denoted v_0, which we assume to correspond to the ambient speed of sound v_1 (in Appendix A we present a more general theory where this is not assumed). We develop the ray model for the curved transducer array; later, we will generalize to other common array types. Beam lines can be considered to spread out from a virtual source (the origin) within the transducer, emanating normal to the transducer surface [36]. The object boundary is represented by the elliptical line $z_{\rm ob}(x)$ which satisfies the relation

$$\frac{x^2}{b^2} + \frac{(z_{\rm ob} - c - d - d_0)^2}{c^2} = 1. \tag{1}$$

where d is the axial distance between the front faces of the object and the transducer, and d_0 is the distance from the virtual source to the transducer surface.

We model the distortion of the object by treating the beam lines as rays. This approximation is valid in the far-field of the source and for objects much larger than the acoustic wavelength. A ray emitted from the transducer surface at point A with coordinates $(x_{\rm A}, z_{\rm A})$ is incident on the front face of the object at point B, $(x_{\rm B}, z_{\rm B})$, as illustrated in figure 1. This follows the straight line path $z_{\rm AB}(x) = x \cot\theta_{\rm A} = (z_{\rm B}/x_{\rm B})x$, where $\theta_{\rm A} = \arctan(x_{\rm B}/z_{\rm B})$ is the angle of the emitted ray relative to the vertical. At B, the angle between the normal of the boundary and

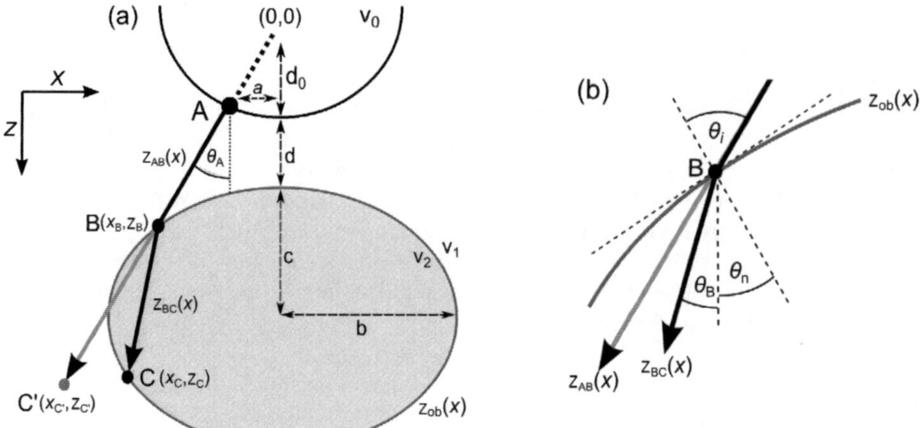

Figure 1. (a) Schematic of the curved transducer array (depicted by the semi-circle at the top of the figure) and elliptical object (speed of sound v_2), embedded within an ambient medium (speed of sound v_1). The ray is emitted from the transducer surface (at point A) at an angle θ_A to the vertical, and is incident at the object surface at point B. For $v_2 \neq v_1$, refraction and misrepresentation of length within the object cause a point C on the far boundary to appear at C'. (b) A close-up of the real ray (black) and image ray (grey) at the point B. The ray is incident on the surface at an angle θ_i to the normal, and the transmitted ray emerges at an angle θ_B to the vertical (with the vertical being at an angle θ_n to the surface normal).

the vertical, θ_n, can be calculated from the gradient of the boundary at B via

$$\tan\theta_\mathrm{n} = \left.\frac{\mathrm{d}z_\mathrm{ob}}{\mathrm{d}x}\right|_{(x_\mathrm{B},z_\mathrm{B})} = \frac{c^2 x_\mathrm{B}}{b^2(d+d_0+c-z_\mathrm{B})}. \tag{2}$$

The angle of incidence at B (relative to the normal) is then $\theta_\mathrm{i} = \theta_\mathrm{n} + \theta_\mathrm{A}$.

At the front face, a component of the wave will be back-scattered to the transducer, registering the presence of this point in the image. Since this ray travels only through the ambient medium, for which the scanner is assumed to be calibrated, the image of this point corresponds to its real location. Of more interest to us is the component of the wave which is refracted through the boundary (the components of the wave that undergo specular reflection or scattering away from the transducer are not registered by the transducer and so we ignore them). This ray is incident on the object's back face at point C, with coordinates $(x_\mathrm{C}, z_\mathrm{C})$, and represented by the straight line $z_\mathrm{BC}(x)$,

$$z_\mathrm{BC}(x) = (x - x_\mathrm{B})\cot\theta_\mathrm{B} + z_\mathrm{B}. \tag{3}$$

θ_B is the angle from the vertical direction, given by $\theta_\mathrm{B} = \theta_\mathrm{r} - \theta_\mathrm{n}$, where θ_r is the angle of refraction. To obtain an expression for θ_r, we rearrange Snell's law, $\sin\theta_\mathrm{i}/\sin\theta_\mathrm{r} = v_1/v_2$, to give

$$\theta_\mathrm{r} = \arcsin\left[\frac{v_2}{v_1}\sin(\theta_\mathrm{A}+\theta_\mathrm{n})\right]. \tag{4}$$

To establish $(x_\mathrm{C}, z_\mathrm{C})$, we seek the intersections of the refracted ray $z_\mathrm{BC}(x)$ with the boundary $z_\mathrm{ob}(x)$. Rearranging equation (3) for x and inserting into equation (1) gives the relation

$$\frac{[(z-z_\mathrm{B})\tan\theta_\mathrm{B} + x_\mathrm{B}]^2}{b^2} + \frac{[z-c-d-d_0]^2}{c^2} = 1. \tag{5}$$

Multiplying this out leads to a quadratic equation in z and thus two solutions for z. The larger of these corresponds to point C (and the smaller to point B). Hence we obtain the z-coordinate of C to be

$$z_{\rm C} = z_{\rm B} + 2\left\{\left[\frac{c^2 x_{\rm B}\tan\theta_{\rm B} - b^2 d' - z_{\rm B} c^2 \tan^2\theta_{\rm B}}{c^2\tan^2\theta_{\rm B} + b^2}\right]^2 + \frac{c^2 b^2 - b^2 d'^2 - c^2\left[x_{\rm B} - z_{\rm B}\tan\theta_{\rm B}\right]^2}{c^2\tan^2\theta_{\rm B} + b^2}\right\}^{1/2} \quad (6)$$

where $d' = d + d_0 + c$. $x_{\rm C}$ then follows from equation (1).

Having established the path to C, we proceed to locate its image, C'. The paths AB and BC have lengths $L_{\rm AB} = \sqrt{x_{\rm B}^2 + z_{\rm B}^2}$ and $L_{\rm BC} = \sqrt{(x_{\rm C} - x_{\rm B})^2 + (z_{\rm C} - z_{\rm B})^2}$, respectively. Assuming that the ray returns to the transducer along the same path, the total travel time is

$$t = \frac{2L_{\rm AB}}{v_1} + \frac{2L_{\rm BC}}{v_2}. \quad (7)$$

The transducer, calibrated to the ambient speed of sound v_1, perceives the ray as the unrefracted path, i.e. along $z_{\rm AB}(x)$, with the same travel time t. Hence the path AC' has length

$$L_{\rm AC'} = \frac{v_1 t}{2} = L_{\rm AB} + \frac{v_1}{v_2} L_{\rm BC}. \quad (8)$$

The position of C' then follows as

$$(x_{\rm C'}, z_{\rm C'}) = (L_{\rm AC'}\sin\theta_{\rm A}, L_{\rm AC'}\cos\theta_{\rm A}). \quad (9)$$

For $v_1 \neq v_2$ the image point differs from the real point, partly due to refraction of the ray at the near face of the object, and secondly due to the rescaling of length due to the speed of sound misrepresentation. For $v_2 < v_1$, the image is extended from the transducer (due to the longer travel time within the object), while for $v_2 > v_1$ the image is moved closer to the transducer (due to the shorter travel time within the object).

2.2. Generalization to Other Array Types

There are 3 other main types of multielement transducer currently in use in medical imaging: the linear, phased, and vector arrays. The ray structure for these array types can be found elsewhere [36]. Below we summarize how the ray model can be extended to these types.

Vector array: In this array, beam lines spread out from a virtual source located within the (flat) transducer. Since we assume that $v_0 = v_1$ (i.e. the scanner calibrated speed of sound equals the ambient speed of sound), no virtual refraction takes place at the transducer surface, and so the only modification is in the position at which the ray emerges from the surface, accounted for in the model by the coordinates $(x_{\rm A}, z_{\rm A})$.

Phased array: In a phased array beam lines spread out from a virtual source at the surface of the transducer, accounted for by setting $d_0 = 0$.

Linear array: In a linear array, beam lines emerge from the transducer surface, i.e. $d_0 = 0$, normal to the transducer face, i.e. $\theta_{\rm A} = 0$. Note that the image position (9) then reduces to

$$(x_{\rm C'}, z_{\rm C'}) = (0, L_{\rm AC'}), \quad (10)$$

i.e. the image point appears directly above the source.

2.3. Area Evaluation

The image boundary will, in general, be irregularly-shaped. To evaluate the enclosed area we exploit Green's theorem [37]. Green's theorem states that, for a region R bound by a simple closed path Γ, then,

$$\iint_R \left(\frac{\partial v}{\partial x} - \frac{\partial u}{\partial z}\right) \mathrm{d}x\ \mathrm{d}z = \oint_\Gamma \{u\ \mathrm{d}x + v\ \mathrm{d}z\}, \tag{11}$$

where $u(x, z)$ and $v(x, z)$ are arbitrary functions, and the path of integration along Γ is taken to be counter-clockwise. Taking the choice of functions $u(x, z) = -z/2$ and $v(x, z) = x/2$ then the left-hand side of equation (11) reduces to $\iint_R \mathrm{d}x\ \mathrm{d}z$, which corresponds to the area of the region R. It then follows that we can express the area A of this region via a line integral around its boundary Γ via

$$\mathrm{A} = \frac{1}{2}\oint_\Gamma (-z\,\mathrm{d}x + x\,\mathrm{d}z). \tag{12}$$

We numerically implement this line integral around the image boundary (discretized as a series of points) to obtain the image area. We quantify the image distortion through the *area distortion* $A_\mathrm{i}/A_\mathrm{r}$, where A_i and A_r are the areas of the image and object. Where discontinuities arise in the image boundary, for the purposes of area evaluation, a straight line traversing the discontinuity is assumed.

3. Materials and Methods

We experimentally analyse the case of a circular object as follows. A hollow spherical aluminium bauble, filled with "liquid 2" and sealed with a latex condom, was suspended (by a wire frame) in a bath of "liquid 1". Its central circular cross-section was imaged from above by an ultrasound scanner (Philips HDI 5000) and curved array (Philips C5-2 40R, operating frequency range 2-5 MHz, d_0 = 40 mm). The set-up is depicted in figure 2(a). The transducer is aligned manually with the equator of the object. The overall object radius r was measured with callipers as r =(3.99 ± 0.01) cm and the transducer-object distance d determined using the scanner's electronic callipers. Where the ambient liquid has the same speed of sound as that assumed by the scanner, this electronic caliper measurement should not suffer any length distortion; where it has a different speed of sound, the electronic caliper measurement is rescaled by a factor v_1/v_0 to account for this distortion. The experiments were run at ambient temperature and the precise bath temperature recorded by digital thermometer.

The aluminium composition of the object was convenient as the high acoustic impedance of aluminium (≈ 17 MRayl) leads to a high contrast image. The aluminium is approximately 1mm thick. At the operating frequency of the scanner probe (2-5 MHz), the wavelength in aluminium is in the range 1-2 mm, and so surface and plate waves can be expected to be highly suppressed. The condom and aluminium shell are sufficiently thin to have negligible impact on our analysis.

For the liquids we employed pure water, pure methanol and an ethanol-water mixture (with 7.25% w/w ethanol). These have room temperature speeds of sound of 1482 m s^{-1} [38], 1100 m s^{-1} [39] and 1540 m s^{-1} [40], respectively (the latter being matched to the tissue average and scanner-assumed value). Note that, while the methanol case of 1100 m s^{-1} is not a clinically relevant speed of sound, for this work it provides a more extreme value with which to further test our ray model. The bath temperature varied in the range 18-23°C and we corrected the speeds of sound used in the model using the established temperature-speed relations [38, 39, 40].

As an initial validation we filled both the object and bath with ethanol-water mixture, such that the system has uniform speed of sound throughout, $v_2 = v_1$ = 1540 m s^{-1}. The B-mode,

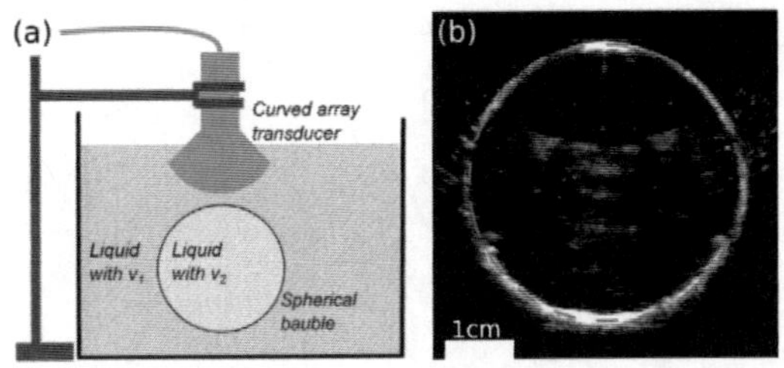

Figure 2. (a) Illustration of the experimental set-up. A curved array transducer, clamped in place, images (from above) the central cross-section of a spherical bauble (diameter 4cm). The bauble is filled with liquid of speed of sound v_2, and the ambient speed of sound is v_1. (b) Corresponding B-mode image for $v_2 = v_1 = 1540$ m s^{-1}, both matched to the scanner. The expected object boundary is shown by the dashed red line.

shown in figure 2(b), shows a circular image boundary, confirming that there is no significant geometric distortion, and matches well to the expected circular shape of the object (red dashed line). Indeed, the experimental boundary (defined as the centre/maximum in the echo) deviates by no more than 1mm from the true object boundary. Note the intensity distortion around the boundary: the poles of the object are brightest as a result of specular reflection.

The area of the object image is estimated using the scanner's built-in "lasso" function. This involves the operator specifying several points around the object boundary (defined as the position of the maximum echo from the boundary) by hand; the scanner then returns the area of the defined region.

Three variations of this system were considered.

$v_2 < v_1$: Taking ethanol-water mixture as liquid 1 and pure water as liquid 2 embodies a system with $v_2 < v_1$ ($v_2 \approx 0.96v_1$).

$v_2 \ll v_1$: Taking the ethanol-water mixture as liquid 1 and now methanol as liquid 2 embodies a system with $v_2 \ll v_1$ ($v_2 \approx 0.71v_1$).

$v_2 > v_1$: Taking pure water as liquid 1 and the ethanol-water mixture as liquid 2 creates a system with $v_2 > v_1$ ($v_2 \approx 1.04v_1$).

Note that, for the first two cases, the speed of sound of the ambient liquid is matched to that assumed by the scanner, as per the assumptions of our ray model. This is not true for the third case, and so the ray model must be extended to account for the distortion of the near-face of the object; this is outlined in Appendix A.

4. Results: Circular Objects

We begin by using the ray model to map out the area distortion of circular objects, first for a simplest case of a linear array and then for our primary case of a curved array. The latter will be validated against experimental images. Unless otherwise stated, we assume the nominal parameters $r = 4$ and $d = 2$. Note that length is subsequently presented in arbitrary units.

4.1. Linear Array

In figure 3(a) we show the ray model prediction for $v_2/v_1 = 0.95$; the typical ray behaviour is illustrated by three sets of rays. The rays propagate vertically downwards from the transducer surface (the x-axis, $z = 0$), impinging the near-face (the upper half) of the object boundary. Our assumption that the scanner is matched to the ambient speed of sound means that the

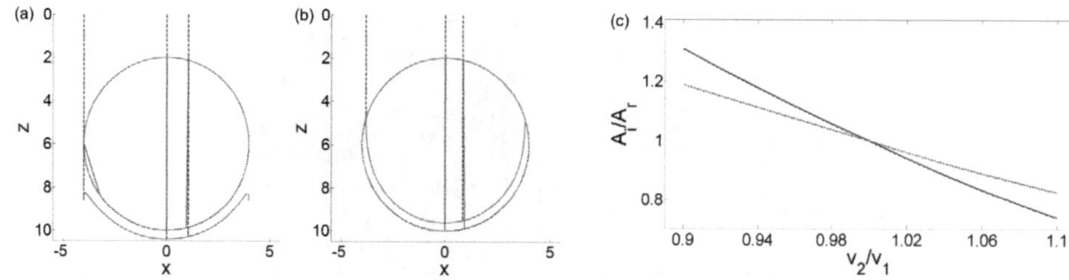

Figure 3. (a) Image (solid green line) of a circular object (dotted black line) with $v_2 = 0.95v_1$ based on a linear array. Three example sets of real rays (dashed blue lines) and their corresponding image rays (solid red lines) are overlaid. (b) Same as above but for $v_2 = 1.05v_1$. (c) Area distortion A_i/A_r as a function of the speed of sound ratio v_2/v_1, for the linear array (green dashed line) and curved array (blue solid line). The curved array case will be further detailed in figure 4.

near face is undistorted, that is, the image boundary (solid green line) and the object boundary (dotted black line) overlap here.

At the near-face the real rays (red lines) are refracted inwards; the exception is the central ray which continues vertically due to its zero angle of incidence, θ_i. These rays travel slower within the object since $v_2 < v_1$. The image rays (dashed blue lines) deviate in two ways: they are unrefracted at the near-face, and they appear elongated within the object. The latter effect is best resolved along $x = 0$: here the image ray is simply extended by a distance $2r(v_1/v_2 - 1) \approx 0.42$. Off-axis, the rays are refracted inwards (towards the normal), with the refraction being small across most of the object since θ_i is small (with the deviation being hard to see by eye). Towards the equator of the object, however, θ_i grows rapidly, causing a significant increase in refraction and the real-ray path length in the object. This has two notable effects as $x \to \pm r$: firstly, the increasing inwards refraction leads to the appearance of a gap in the image boundary, and secondly, the divergence of the internal path length leads to the formation of prominent "flicks" in the image boundary [14].

Now consider the case $v_2/v_1 = 1.05$, shown in figure 3(b). Since $v_2 > v_1$ the real rays travel faster than the image rays within the object, and so the far image boundary becomes shifted towards the transducer. Also, the real rays become refracted outwards (away from the normal). This removes the appearance of the above flicks for $x \approx \pm r$. The image boundary again features a gap, although its origin is different to that above: here it is caused when the rays impinge the near face at a sufficiently large angle of incidence that they undergo total internal reflection, and so do not reach the far face.

For the above cases the area distortion is $A_i/A_r = 1.093$ and 0.907, respectively. In other words, a 5% speed of sound difference has led to $\approx 10\%$ error in the area. We have also explored the wider dependencies of the area distortion on the key parameters. The area distortion scales approximately linearly with v_2/v_1, as shown in figure 3(c). As d, the transducer-object distance, is varied, the image area A_i is unaffected, since all rays approach the object in the same direction. Moreover, the image area scales in proportion to the size of the circular object, such that the area distortion ratio A_i/A_r is independent of r, the radius of the object.

4.2. Curved Array

The corresponding images for a curved array are shown in figure 4(a-b). We see similar qualitative effects as for the linear array: the shifting of the image boundary in front/behind the object boundary, gaps in the image boundary, and the appearance of flicks for $v_2 < v_1$.

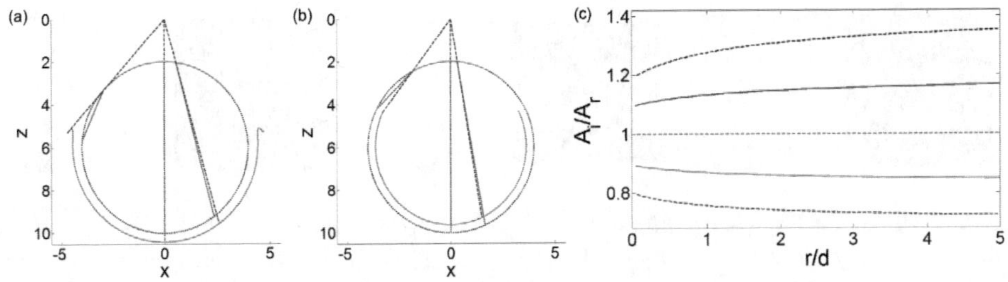

Figure 4. (a)-(b) As figure 3 for a curved array (with $d + d_0 = 2$). (c) Area distortion A_i/A_r of the circular object under curved array imaging as a function of the ratio r/d (recall, r is the object radius and d the distance from the object near-surface to the transducer). Various values of v_2/v_1 are considered ranging from 0.9 (top line) to 1.1 (lowest line) in steps of 0.05. The values $v_2/v_1 = 0.95$ and 1.05 are highlighted by solid lines.

However, the ray structure and image shape are modified further due to the angular spread of the emitted rays. The near-face of the object has reduced extent, such that a greater proportion of the boundary is distorted. The incident rays now approach the tangent of the object fore of the equator, which causes a corresponding shift in the gaps and flicks in the boundary towards the transducer. For the $v_2 < v_1$ case we find $A_i/A_r = 1.147$; for $v_2 > v_1$ we find $A_i/A_r = 0.859$. In other words, a 5% difference in sound speeds for the curved array leads to over 14% deviation in area.

As for the linear array, the variation of the area distortion with v_2/v_1 is approximately linear [figure 3(c)], but with greater deviation than for the linear array. The dependence on r and d is more complicated for the curved array since these parameters control the angular spread of the incident rays. Figure 4(c) shows that the change in area distortion (relative to the undistorted value $A_i/A_r = 1$) increases with r/d; this is due to the increasing angular spread of rays as r/d increases, leading to greater refractive effects and increasing the proportion of the boundary which is distorted.

4.3. Experimental Validation

Experimental B-mode images (with curved array) of the circular test object are shown in figure 5(i), alongside (ii) the ray model predictions of the distorted boundary, for following 3 speed of sound scenarios.

(a) $v_2 < v_1$: Here we consider $v_1 = 1540$ m s^{-1} (ethanol-water mixture) and $v_2 = 1490$ m s^{-1} (water), giving $v_2/v_1 = 0.96$. No geometric distortion is evident at the near face but the far face is shifted away from the transducer, in good agreement with the ray model (dashed green line). The predicted area distortion is $A_i/A_r = 1.076$. The scanner area estimate is $A_i/A_r = 1.074$, in good agreement with the theoretical result.

(b) $v_2 \ll v_1$: Now taking $v_2 = 1100$ m s^{-1} (methanol) gives the more extreme case of $v_2/v_1 = 0.71$. The far-face of the image is highly distorted, in good agreement with the ray model. The theoretical area distortion is $A_i/A_r = 1.97$ (i.e. the image has approximately twice the area of the object), close to the scanner estimate $A_i/A_r = 2.02$.

(c) $v_2 > v_1$: Lastly, we take $v_1 = 1490$ m s^{-1} (water) and $v_2 = 1540$ m s^{-1} (ethanol-water mixture), giving $v_2/v_1 = 1.04$. The ambient liquid is no longer matched to the scanner calibration, causing the near face to become slightly shifted *towards* the far face. Again, the overall image shape agrees well with the ray model (in this case based on the extended

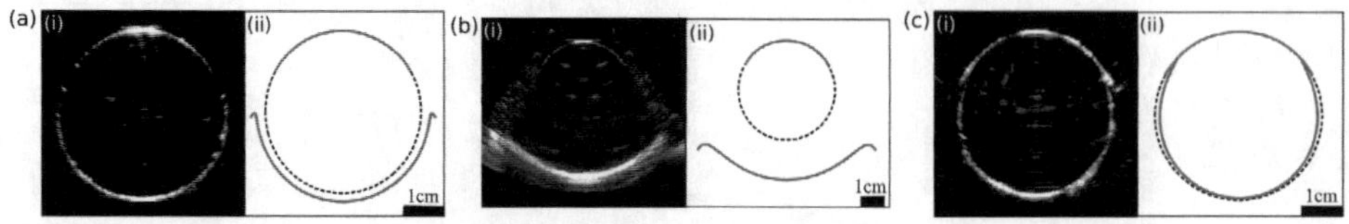

Figure 5. (i) B-mode images of the circular object of speed v_2 (imaged from above in a liquid with speed v_1), for various speed of sound combinations, with (ii) the corresponding ray model prediction of the image boundary (green lines). In (ii) the object boundary is also shown (black dashed line). (a) $v_2 = 1490$ m s^{-1} and $v_1 = 1540$ m s^{-1}, giving $v_2/v_1 = 0.968$. (b) $v_2 = 1100$ m s^{-1} and $v_1 = 1540$ m s^{-1}, giving $v_2/v_1 = 0.71$. (c) $v_1 = 1490$ m s^{-1} and $v_2 = 1540$ m s^{-1}, giving $v_2/v_1 = 1.034$.

theory in Appendix A). The theoretical area distortion $A_i/A_r = 0.945$ is close to the scanner estimate $A_i/A_r = 0.966$.

These results validate the use of the ray model in predicting the geometric distortion of the object boundary.

5. Results: Elliptical Objects

We extend our theoretical analysis to the more general case of elliptical objects, which allows us to explore the effect of elongation. For brevity, we focus on the curved array.

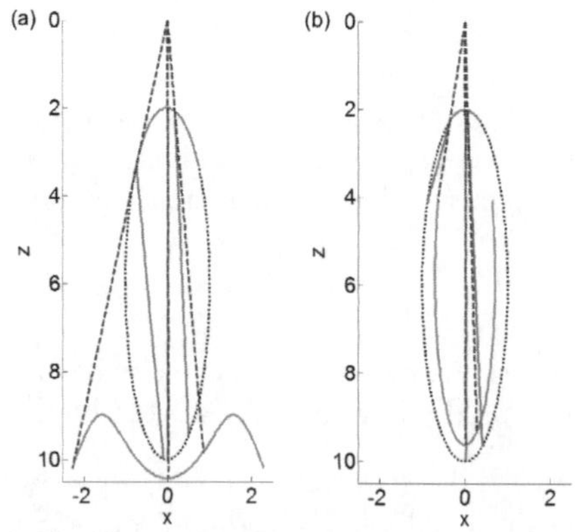

Figure 6. Curved array images (solid green line) of an elongated elliptical object (dotted black line) with $b = 1$ and $c = 4$ ($d + d_0 = 2$). Shown are the results for (a) $v_2 = 0.95v_1$ and (b) $v_2 = 1.05v_1$.

We first consider an elliptical object that is elongated along z, with nominal parameters $b = 1$ and $c = 4$. The image distortion, presented in figure 6 for both (a) $v_2 = 0.95v_1$ and (b) $v_2 = 1.05v_1$, shows similar qualitative structure for the circular object [figure 4] but with enhanced distortion. For $v_2 < v_1$, the flicks are even more pronounced than for the circular case, caused by the fast roll-off of the object boundary with x (which causes the angle of incidence to increase more quickly with x). The area distortions are $A_i/A_r = 1.58$ (for $v_2 < v_1$) and 0.68 (for $v_2/v_1 > 1$); this is vastly greater than for circular objects in Section 4.2, demonstrating the sensitive role of elongation.

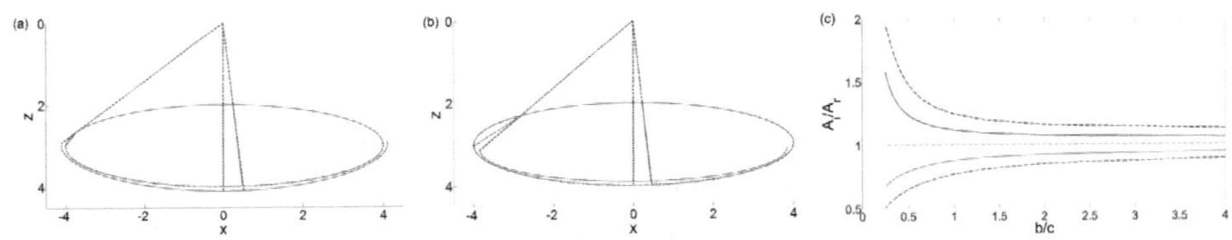

Figure 7. (a)-(b) Same as figure 6 but for a flattened elliptical object with $b = 4$ and $c = 1$. (c) Area distortion A_i/A_r versus ellipticity b/c (with b fixed to unity) under curved array imaging. Various values of v_2/v_1 are considered ranging from 0.9 (top line) to 1.1 (lowest line) in steps of 0.05.

In contrast, for a flattened ellipse [figure 7 (a)-(b), nominal parameters $b = 4$ and $c = 1$] the image distortion is reduced, with negligible flicks appearing in the image due to the slower roll-off of the boundary with x. Also, the narrowness of the object in z means that the rays have little distance to accumulate any significant distortion. The area distortions here are $A_i/A_r = 1.06$ (for $v_2/v_1 < 1$) and 0.95 (for $v_2/v_1 > 1$), confirming the reduced distortion for flattened objects.

To further quantify the dependence on elongation, figure 7 (c) plots the area distortion A_i/A_r as a function of ellipticity b/c. We clearly see that the distortion diverges for objects which are increasingly elongated (along z), and converges towards the undistorted result for increasingly flattened geometries.

We have compared the ray model predictions against experimental images of an elliptical object. These results, presented in Appendix B, further support the validity of the ray model in modelling the distortion of the boundary in the image.

6. Discussion and Conclusions

We have developed a ray model that describes the geometric distortion of two-dimensional ultrasound images (based on linear, curved, phased or vector arrays) of circular and elliptical objects due to speed of sound discrepancies. The geometric distortion arises from the refraction of the sound and the misrepresentation of length within the object. By comparing to ultrasound images of a test object we have validated that the ray model successfully captures the underlying geometric distortion of the image.

The geometric distortion of ultrasound images of circular objects has been examined previously [15, 16], including by ray models [14, 13]. Our analysis extends this previous body of work by *quantifying* the distortion of area, and thereby providing new insight into the errors in the ultrasonic evaluation of area and volume of anatomical features. Our work also extends previous analyses by considering elliptical objects, which allows us to explore the role of elongation, and by covering all four of the main transducers array formats.

When $v_2 < v_1$, the image is larger than the object, and vice versa for $v_2 > v_1$. The ray model predicts the emergence of gaps in the image boundary, an effect which may contribute to the phenomenon of acoustic shadowing [14, 15, 16, 13]. For $v_2 < v_1$ the image also features distinctive flicks at the extreme angular positions, as noted previously [14]. The deviation of the image area from the true value is greater for the "angled" array systems (phased, curved and vector arrays) than the linear array, due to the enhanced role of refraction. The deviation also depends on the shape and size of the object, in general increasing with the lateral size of the object and its elongation along the ultrasound axis.

In Ref. [16] it is argued that the true dimension of the object, along the axis of the interrogating beam, is the sonographic dimension of the object multiplied by the ratio of the object speed of sound to the scanner-assumed speed of sound (v_2/v_1 in our notation). In other words, the true object shape should correspond to the image shape scaled axially by a factor

v_2/v_1. From this it follows that the true and image areas are related by $A_r/A_i = v_2/v_1$. This linear relationship is shown in figure 8 (red line). However, we typically find the distortion to be considerably larger than this prediction, a difference which is then clearly attributable to refraction (not included in this result from Ref. [16]). To illustrate this, we have superimposed onto this figure our results (same parameters as figure 3) for a circular object based on a linear (green line) and curved array (blue line). It is evident that the Ziskin prediction underestimates the distortion, typically by a factor of 2-3 for these examples (more generally this factor will depend on the object shape, size and distance from the transducer). For example, for $v_2/v_1 = 1.05$, the ray model predicts $A_r/A_i = 1.1$ for a linear array and 1.16 for a curved array (c.f. 1.05 according to [16]). The area distortion has contributions from refraction of the rays at the near surface and the miscalculation of the length of the ray within the object. We can artificially remove the effect of refraction from the ray model by setting $\theta_r = \theta_i$. In doing so we find for a circular object that these effects roughly contribute to of the order of half of the area distortion (although, again, this will depend in general on the object parameters and shape).

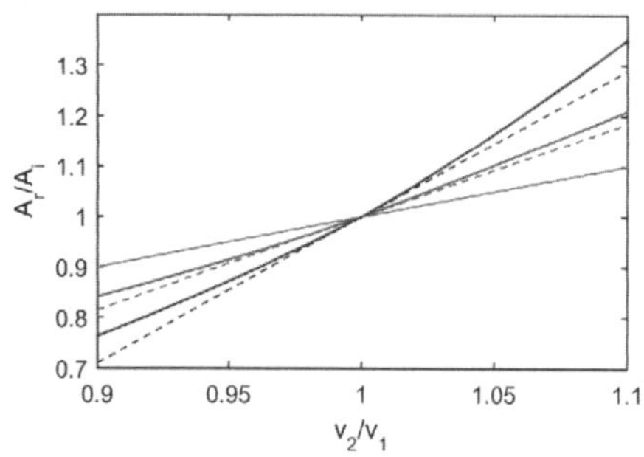

Figure 8. Ratio of the true object area to the image area, A_r/A_i, versus the speed ratio v_2/v_1. The ray model predictions for a circular object imaged via linear and curved array are shown by green and blue lines, respectively (the model parameters are as in figure 3). The prediction $A_r/A_i = v_2/v_1$, derived from Ref. [16], is shown by the red line. As a guide to the eye, straight lines with gradient 1.85 and 2.9 are shown by the dashed lines.

In its conventional clinical use, ultrasound provides qualitative, diagnostic images within the body. Increasingly, however, medical ultrasound is being pursued to provide *quantitative* volumetric measurements of organs and tumours [20, 21, 22]. This functionality allows, for example, the precise mapping of tumour growth and the efficacy of therapies, as well as the non-invasive measurement of urine volume in the bladder (negating the need for conventional cathetization techniques, which are uncomfortable and present an infection risk). Following improvements in 3D reconstruction of ultrasound images [34], ultrasound-based volume measurements are claiming to have errors of only a few percent across a variety of anatomical structures [26, 22, 29, 28]. These measurements do not take into account the effects of variable speed of sound which will geometrically distort the image and lead to systematic errors in the volume measurement. With 3D reconstructed images typically being formed from 2D images, the problem is essentially a 2D one. As such, these geometric distortions we examine here pose a key limiting factor on the current possible accuracy of these measurements. We hope to examine the 3D distortions in future work. Moreover, developments in this direction could lead to algorithms to correct the distortion in clinical measurements, allowing area and volume measurements which overcome the distortion limit.

Appendix A. Ray Model for $v_1 \neq v_0$

For simplicity our main ray model considered the ambient speed of sound v_1 to be matched to that of the scanner v_0, thereby negating the misrepresentation of distance as the ray travels through the ambient liquid. Here we extend the ray model to account for $v_1 \neq v_0$ with a curved array.

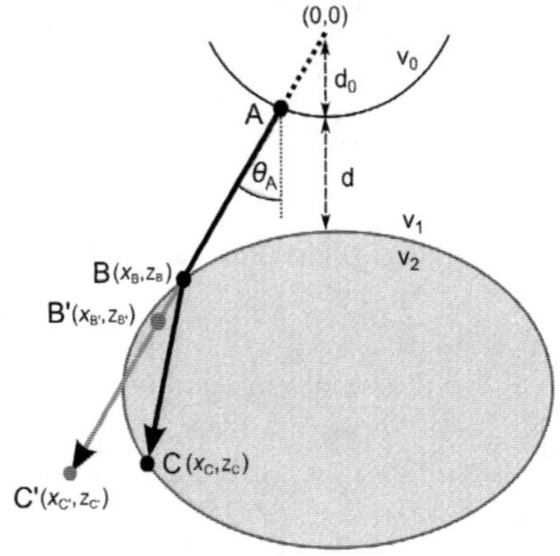

Figure A1. Schematic of the transducer-object scenario, featuring an elliptical object (speed of sound v_2) within an ambient medium (speed of sound v_1). The transducer is calibrated to a speed of sound v_0.

We consider a transducer element on the surface of the curved array, point A [figure A1]. Since the ray is emitted *normal* to the surface, no refraction effects occur at the transducer boundary, despite $v_1 \neq v_0$. The path of the real ray, which refracts into the object at point B on the near face and reaches the far boundary at point C, is provided by the same theory as before. The difference, however, now arises in the position of point B and its image point B'. We add a misregistration to all of the points by introducing the two relations for the misplacement of x and z, namely:

$$\Delta X = \left(\frac{v_0}{v_1} - 1\right) \left[\sqrt{x^2 + z^2} - d_0\right] \sin\theta_\mathrm{A},$$

$$\Delta Z = \left(\frac{v_0}{v_1} - 1\right) \left[\sqrt{x^2 + z^2} - d_0\right] \cos\theta_\mathrm{A},$$

which are similar to those in Ref. [36]. Thus the position of the image point B' is

$$(x_{\mathrm{B'}}, z_{\mathrm{B'}}) \quad = \quad (x_\mathrm{B} + \Delta X, z_\mathrm{B} + \Delta Z).$$

The total length of the virtual ray, the equivalent of equation (8), is

$$L_{\mathrm{AC'}} = \frac{v_0}{v_1} L_{\mathrm{AB}} + \frac{v_0}{v_2} L_{\mathrm{BC}},$$

and the image point C' is located at

$$(x_{\mathrm{C'}}, z_{\mathrm{C'}}) = (L_{\mathrm{AC'}} \sin\theta_\mathrm{A}, L_{\mathrm{AC'}} \cos\theta_\mathrm{A}).$$

Thus the distortion of the near-face and the far-face is completely determined.

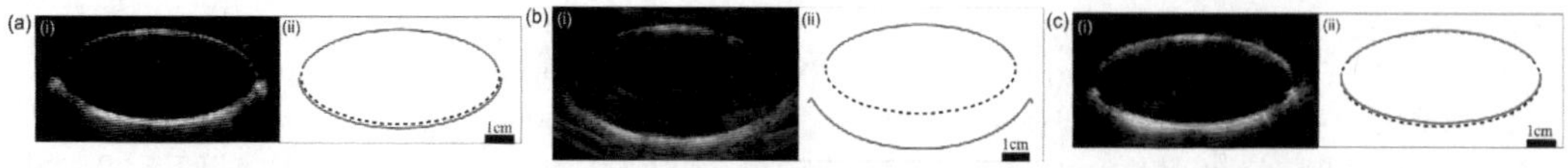

Figure B1. (i) B-mode images of the elliptical 2D object with $b/c \approx 1.9$ for various speed of sound combinations, with (ii) the corresponding ray model prediction of the image boundary (green lines). In (ii) the object boundary is also shown (black dashed line). (a) $v_2 = 1490$ m s^{-1} and $v_1 = 1540$ m s^{-1}, giving $v_2/v_1 = 0.968$. (b) $v_2 = 1100$ m s^{-1} and $v_1 = 1540$ m s^{-1}, giving $v_2/v_1 = 0.71$. (c) $v_1 = 1490$ m s^{-1} and $v_2 = 1540$ m s^{-1}, giving $v_2/v_1 = 1.034$.

Appendix B. Experimental images of elliptical objects and comparison to the ray model

As well as comparing against experimental images of a circular object, presented in Section 4.3, we have further compared against images of elliptical geometry. We repeat the methodology described in Section 3 using a hollow ellipsoidal shell. The shell is made of acrylic plastic, 1.5 mm thick. The shell is oblate (flattened along its axis of rotational symmetry), with external dimensions 31.8 mm (short diameter)) by 69.4 mm (long diameter). When the axis is aligned vertically and imaged directly from above, the imaged cross-section is a flattened ellipse with $b/c \approx 1.9$; when the axis is aligned horizontally, the imaged cross-section is an elongated ellipse with $b/c \approx 0.5$. The images for these two shapes for the three speed of sound cases (detailed in Section 3) are shown in Fig. B1 and Fig. B2, respectively.

The image boundary is of lower contrast and less sharp than for the circular object (Fig. 5), due to two differences. Firstly, the acoustic impedance of acrylic (≈ 3.3 MRayl) is approximately 5 times less than that of aluminium (≈ 17 MRayl) [41], giving a reduced reflectivity from the elliptical object. Secondly, the thicker shell of the elliptical object leads to less sharp boundary and some occurrences of reverberation echoes (see, e.g. the lower face of Fig. B1(c)(i)).

For the flattened case (Fig. B1) the distortion is evident for all three speed of sound scenarios, that is, the shift of the far face, the discontinuity in the image towards the equator of the object, and the distortion from an elliptical shape. This departure is particularly large for the case $v_2 \ll v_1$ (Fig. B1(b)), where the far face is appears at significantly increased depth, and the far face is stretched transversely by refractive effects, in good agreement with the prediction of the ray model. For three speed of sound cases (a-c), the model predicts $A_i/A_r = 1.076$, 1.63 and 0.98, which are close to the scanner estimates of $A_i/A_r = 1.05, 1.55$ and 0.97.

For the elongated case (Fig. B2), the image quality of the far (lower) face is further reduced. This is primarily due to the prominence of oblique rays (incident close to the object equator where the boundary is approximately vertical) which spend considerable time in the shell, introducing artefacts due to propagation through the shell itself (not accounted for within the model). To a lesser extent, the image quality of the far face is also reduced due to its relatively large distance from the transducer. Nonetheless, there is some qualitative agreement with the ray model, including the vastly shifted and stretched lower face for case (b) and the "squashing" of the lower face in case (c).

References

[1] Rumack C, Wilson S, Charboneau J and Levine D 2011 *Diagnostic Ultrasound 4th Edition* (Elsevier Mosby, Philadelphia)
[2] Fontanarosa D, Pesente S, Pascoli F, Ermacora D, Rumeileh I A and Verhaegen F 2013 *Phys. Med. Biol.* **58** 1341
[3] Duck F 2012, *Physical Properties of Tissue* (Institute of Physics and Engineering in Medicine, York)
[4] O'Donnell M and Flax S W 1988 *Ultrason. Imaging* **10** 1
[5] Anderson M E, McKeag M S and Trahey G E 2000 *J. Acoust. Soc. Am.* **107** 3540

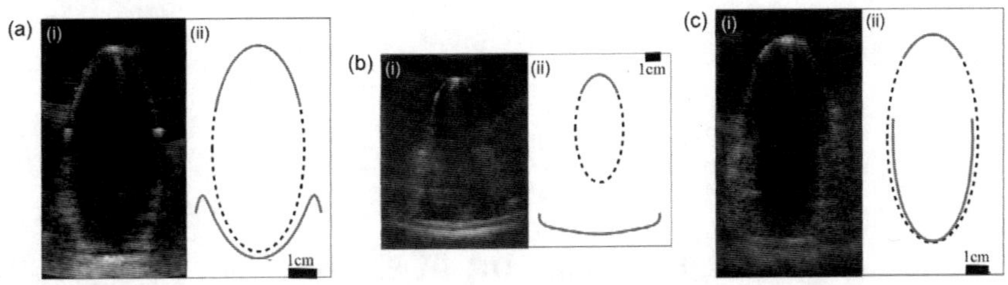

Figure B2. As Fig. B1 but when the object is rotated by 90°, realizing a 2D elliptical object with $b/c \approx 0.5$.

[6] T L Szabo 2004 *Diagnostic Ultrasound Imaging: Inside Out* (Elsevier, London)
[7] Napolitano D, Chou C-H, McLaughlin G, Ji T-L, Mo L, DeBusschere D and Steins R 2006 *Ultrasonics* **44** e43
[8] Cho M H, Kang L H, Kim J S and Lee S Y 2009 *Ultrasonics* **49** 774
[9] Shin H-C, Prager R, Gomersall H, Kingsbury N, Treece G and Gee A 2010 *Ultrasonics* **50** 716
[10] Yoon C, Lee Y, Chang J H, Song T and Yoo Y 2011 *Ultrasonics* **51**, 795
[11] Jaeger M, Robinson E, Akarcay H G and Frenz M 2015 *Phys. Med. Biol.* **60** 4497
[12] Robinson D E, Kossof G and Garrett W J 1966 *Ultrasonics* **4** 186
[13] Sommer F, Filly R and Minton M 1979 *Am. J. Roentgenol.* **132** 973
[14] Robinson D, Wilson L, Kossoff G 1981 *J. Clin. Ultrasound* **9** 181
[15] LaFollette P and Ziskin M 1986 *Ultrasound Med. Biol.* **12** 953
[16] Ziskin M, LaFollette P, Blathras K and V Abraham 1990 *Ultrasound Med. Biol.* **1** 183
[17] Opielinski K J and Gudra T 2000 *Ultrasonics* **38** 424
[18] Rubin J, Adler R, Fowlkes J and Carson P 1991 *Ultrasound Med. Biol.* **17** 85
[19] Steel R, Poepping T, Thompson R and Macaskill C 2004 *Ultrasound Med. Biol.* **30** 1153
[20] Campani R, Bottinelli O, Calliada F and Coscia D 1998 *Eur. J. Rad.* **27** S183
[21] Gilja O H, Hausken T, Berstad A and Odegaard, S 1999 *Proc. Instn Mech. Eng. H.* **213** 247
[22] Treece G, Prager R, Gee A and Berman L 2001 *Med. Im. Anal.* **5** 41
[23] Fontanarosa D, van der Meer S, Harris E and Verhaegen F 2011 *Medical Physics* **38** 2665
[24] Riccabona M, Nelson T and Pretorius D 1996 *Ultrasound Obstet. Gynecol.* **7** 429
[25] Riccabona M, Nelson T and Pretorius D 1996 *J. Ultrasound Med.* **15** 627
[26] Kristiansen N, Djurhuus J and Nygaard H 2004 *Med. Biol. Eng. Comput.* **42** 762
[27] Suwanrath C, Suntharasaj T and Sirapatanapipat H 2009 *J. Ultrasound Med.* **28** 847
[28] Tong S, Downey D B, Cardinal H N and Fenster A 1996 *Ultrasound Med. Biol.* **22** 735
[29] Huang Q H, Zheng Y P, Lu M H and Chi Z R 2005 *Ultrasonics* **43** 153
[30] Yoshizaki C T, Francisco R P V, de Pinho J C, Ruano R and Zugaib M 2013 *J. Ultrasound Med.* **32** 421
[31] Lyshchik A, Drozd V, Schloegl S and Reiners C 2004 *J. Ultrasound Med.* **23** 247
[32] Hornblower V D M, Yu E, Fenster A, Battista J J and Malthaner R A 2007 *Phys. Med. Biol.* **52** 91
[33] Linney A D and Deng J 1999 *Proc. Instn Mech. Eng. H.* **213** 235
[34] Fenster A, Downey D B and Cardinal H N 2001 *Phys. Med. Biol.* **46** R67
[35] Betrouni N, Lopes R, Makni N, Dewalle A S, Vermandel M and Rousseau J 2009 *Ultrasonics* **49** 646
[36] Goldstein A 2000 *Ultrasound Med. Biol.* **26** 1133
[37] Stewart J 2012 *Calculus 7th Edition* (Brooks/Cole, Belmont, USA)
[38] Bilaniuk N and Wong G 1993 *J. Acoust. Soc. Am.* **93** 1609-1612
[39] Benson G 1979 *J. Chem. Thermodyn.* **11** 861
[40] Tong J, Povey M, Zou X, Ward B and Oates C 2011 *J. Phys: Conf. Ser.* **279** 012023
[41] *Tables of Physical and Chemical Constants* (16th edition 1995). Kaye and Laby Online (www.kayelaby.npl.co.uk). Version 1.0 (2005)

Microultrasound characterisation of *ex vivo* porcine tissue for ultrasound capsule endoscopy

H S Lay[1], B F Cox[2], M Sunoqrot[2], C E M Démoré[3], I Näthke[2], T Gomez[4], S Cochran[1]

[1]University of Glasgow, Glasgow, UK
[2]University of Dundee, Dundee, UK
[3]Medical Biophysics, University of Toronto, Toronto, Canada
[4]Spanish National Research Council, Madrid, Spain
holly.lay@glasgow.ac.uk

Abstract Gastrointestinal (GI) disease development and progression is often characterised by cellular and tissue architectural changes within the mucosa and sub-mucosa layers. Current clinical capsule endoscopy and other approaches are heavily reliant on optical techniques which cannot detect disease progression below the surface layer of the tissue. To enhance the ability of clinicians to detect cellular changes earlier and more confidently, both quantitative and qualitative microultrasound (μUS) techniques are investigated in healthy *ex vivo* porcine GI tissue. This work is based on the use of single-element, focussed μUS transducers made with micromoulded piezocomposite operating at around 48 MHz.

To explore the possibility that μUS can detect Crohn's disease and other inflammatory bowel diseases, *ex vivo* porcine small bowel tissue samples were cannulised and perfused with phosphate-buffered saline followed by various dilutions of polystyrene microspheres. Comparison with fluorescent imaging showed that the microspheres had infiltrated the microvasculature of the samples and that μUS was able to successfully detect this as a mimic of inflammation. Samples without microspheres were analysed using quantitative ultrasound to assess mechanical properties. Attenuation coefficients of 1.78 ± 0.66 dB/mm and 1.92 ± 0.77 dB/mm were obtained from reference samples which were surgically separated from the muscle layer. Six intact samples were segmented using a software algorithm and the acoustic impedance, *Z*, for varying tissue thicknesses, and backscattering coefficient, BSC, were calculated using the reference attenuation values and tabulated.

1. Introduction

Ultrasound endoscopy represents the state-of-the-art in clinical diagnostic tools for diseases of the human gastrointestinal (GI) tract. High-definition optical imaging is used in conjunction with ultrasonic imaging to visualise the surface and penetrate the thickness of the tissue to identify the cellular and tissue architectural changes induced by disease. This technology is limited, however, by the length of the endoscope probe, rendering the full extent of the GI tract difficult to visualize properly.

Video capsule endoscopy (VCE) is a technology which is just now reaching maturity after two decades of development, using physical implementation as a capsule to eliminate the probe length limitations of standard endoscopes and allowing visual diagnosis of the full length of the GI tract [1]. However, current capsules can perform only optical imaging, limiting their diagnostic capability. This has motivated research into expanding the diagnostic modalities available in capsule format [2], including research seeking to develop ultrasound capsule endoscopy (USCE) integrating microultrasound (μUS) imaging systems with high-quality optics and additional modalities.

Coeliac disease and inflammatory bowel disease (IBD) such as Crohn's disease represent common clinically encountered gastrointestinal disorders that significantly affect the health and well-being of patients. VCE has been indicated in the management of both conditions because of its ease of use and acceptability to patients as compared to standard endoscopy [3]. However, the lack of subsurface imaging available in VCE limits its usefulness in the diagnosis of early Coeliac or Crohn's disease, as both disorders originate below the surface of the mucosa [4]. For this reason, the functionality of μUS imaging, above the standard range of 1-20 MHz, merits investigation in examining the layer structure of healthy GI tissue and modelling subsurface inflammation similar to that seen in Crohn's and Coeliac disease.

It has already been shown that μUS can be used to characterize GI tissue with a high level of agreement with histology [5]. This paper first reports qualitative studies of healthy tissue and tissue treated with microspheres to mimic inflammation. In addition to qualitative analysis, μUS can also be used to determine the mechanical properties of tissue based on its effects on acoustic wave propagation and related investigations have shown that quantitative μUS can be used to detect pre-cancerous changes in tissue structures [6]–[9]. Two measures which have been identified as of interest to this study are the acoustic impedance (Z) and backscattering coefficient (BSC), which have shown potential for detecting cellular and tissue architectural changes characteristic of early disease stages [6].

A disease whose progression includes the types of cellular change that may be addressed is Barrett's Esophagus (BE). In BE, the distal squamous mucosal layer of the esophagus undergoes a metaplastic transformation and is replaced by a simple columnar layer. BE affects approximately 2% of the UK population [10] and is considered a pre-malignant condition at risk of developing into oesophageal cancer. For this reason, oesophageal tissue samples were targeted for quantitative analysis, focusing on measurements of the type that could be performed *in vivo* to allow Z and BSC to be calculated.

In this paper, *ex vivo* porcine tissues were used in both the qualitative and the quantitative studies to establish a healthy baseline and to establish resolution thresholds in the measurements proposed. Emphasis was placed on measurement techniques which could be translated to *in vivo* clinical implementation with USCE with a minimum of change, so calibration scans are used for those measurements which could not be performed without a ground truth reference.

Healthy swine oesophageal and small bowel tissue was chosen primarily because the histology of the porcine gastrointestinal tract resembles that of humans at the squamocolumnar junction [11]. As tissues were obtained fresh-frozen from the abattoir post-mortem, ethical approval was not necessary. While the frozen state of the tissue was a consideration, freeze-thaw cycles have been shown not to affect the acoustic properties of tissue, allowing them to be used in the analysis without concern [12].

2. Tissue Characterisation

2.1. Qualitative Imaging

The feasibility of an ultrasound-based endoscopic capsule system will be dependent on the ability to distinguish health-related changes in the subsurface structure of the GI tract. To investigate this, we generated single-element microultrasound transducer scans of healthy porcine GI tissue and examined them qualitatively to establish base-line values against which diseased tissue could be assessed. To mimic the tissue inflammation seen in Crohn's disease and Celiac disease, fluorescent microspheres were infused into cannulised bowel tissue and imaged both optically and with μUS. The fluorescent properties of the microspheres allowed them to be visualised optically while their acoustic properties were assessed in the μUS images.

2.2. *Quantitative Measures*

2.2.1. *Attenuation*

As the tissue samples analysed for this paper were of non-negligible thickness, it was critical to assess and compensate for the loss-factors which could skew the raw echo data and bias the resulting values. Some of these factors cannot be measured in *in vivo* tissue because of a lack of ground truth references, so typical values are calculated from *ex vivo* tissue samples with similar properties using the same scanning set up. By placing the *ex vivo* tissues on top of a highly-reflective substrate and imaging the substrate both with and without the intervening tissue, the two-way attenuation of the tissue can be measured. While attenuation in the tissue is caused by both absorption and scattering, which are frequency dependent, a bulk attenuation factor can be calculated which will be consistent for tissue of a specific type and similar thickness scanned with the same centre frequency.

For a tissue sample with known thickness d, the attenuation factor (α) in dB mm^{-1} can be calculated as:

$$\alpha = \frac{-20}{2*d}\log_{10}\frac{V_r}{V_a}, \tag{1}$$

where V_r is the amplitude of the reflected signal received from the reference reflector beneath the tissue and V_a is the amplitude of the signal reflected from the reference reflector with no masking tissue. The logarithmic attenuation coefficient (dB mm^{-1}) can then be converted to the natural logarithmic a' (neper mm^{-1}) via:

$$a' = \frac{\alpha}{8.686}. \tag{2}$$

2.2.2. *Acoustic Impedance*

The characteristic acoustic impedance of a tissue, Z, is a function of the density and wave velocity of a tissue, and thus can be used to determine if changes in the overall density have occurred because of re-organization of the cellular structure or changes in the fluid levels. If the tissue of interest is embedded in a fluid of known impedance, the easiest method to calculate the impedance is through analysis of the echo reflection as a function of the reflection from a known material. For both the known and unknown materials, the reflected wave amplitude (V_r) is a function of the incident wave amplitude (V_i) and the acoustic impedances of the two media,

$$V_r = \frac{Z_t - Z_w}{Z_t + Z_w} V_i, \tag{3}$$

where Z_t is the acoustic impedance of the material under investigation and Z_w is the acoustic impedance of the imaging medium, given a perpendicular incident wave and $Z_t > Z_w$. Rearranging for the unknown target acoustic impedance results in a solution for Z_t (MRayl):

$$Z_t = -\frac{V_r + V_i}{V_r - V_i} Z_w. \tag{4}$$

The known reflector is used to solve for V_i using the inverted version of eqn 3:

$$V_i = \frac{1}{R_q} V_r, \tag{5}$$

where R_q is the reflection coefficient defined by:

$$R_q = \frac{Z_t - Z_w}{Z_t + Z_w}. \tag{6}$$

Solving this equation for the reference reflector (a quartz flat with known material properties) results in a value for V_i, which can then substituted into Eqn. 4 to solve for Z_t for the unknown material.

2.2.3. Backscatter Coefficient
Calculating the BSC of a material can provide insight into the distribution and properties of scatterers which are smaller than the imaging wavelength. These would be otherwise invisible to normal ultrasound imaging. Because of the dependence of backscatter on the scanning procedure and transducer properties, the calculations in this paper are based on the technique derived by Foster et al [13] for use with single-element focused transducers. Using this derivation, E_S, the energy received by the transducer after scattering by the tissue can be written as:

$$E_s = KI_0A \int_S \mu(\theta)\, d\Omega \int_{d_1}^{d_2} e^{-4a'l} dl, \tag{7}$$

where K is a constant propagation factor, I_0 is the incident acoustic intensity from the transducer, and A is the surface area of the transducer face. The differential scattering coefficient, $\mu(\theta)$, is integrated over the surface of the receiver, S, and the attenuation coefficient, a', is integrated from d_1, the minimum depth of the tissue window, to d_2, the maximum depth of the tissue window [13].

Breaking up Eqn. 7 to isolate terms, it can be shown that the first term is dependent on the scanning transducer characteristics and the last term is a function of the attenuation properties of the material. Using a known ground truth reference similar to that used in the acoustic impedance measurements, the first term can be cancelled out by obtaining a reference scan with the same scanning setup as the tissue scans. By substituting in reference scan values and rearranging results, it can be shown [13] that the backscatter coefficient ($Sr^{-1}mm^{-1}$) is:

$$\mu_B = \frac{R_q}{2\pi(1-\cos\theta_T)} \frac{\int_{t_1}^{t_2} |V_S(t)|^2 dt}{\int_{-\infty}^{\infty} |V_q|^2 dt} \frac{4a'}{e^{-4a'd_1} - e^{-4a'd_2}}, \tag{8}$$

where θ_T is the half-angle subtended by the transducer face at the focal point, V_s is the signal obtained from the interrogated sample, integrated over the time-gate of interest, and V_q is the amplitude of the signal obtained from the ground truth reference, integrated over the full sample time.

3. Tissue Preparation

3.1. General Tissue Preparation
Porcine tissue from the oesophagus and small bowel were obtained frozen from an abattoir (Medical Meat Supplies Ltd, Oldham, UK) and the tissue samples were stored in a freezer at -4°C. These samples were assumed to be of healthy tissue as they were originally intended for human consumption. The oesophageal samples were obtained with the stomach still attached to insure presence of the gastroesophageal junction (GOJ), while the bowel samples had the mesentery and mesenteric vessels intact for perfusion purposes.

All tissue samples were initially thawed under running tap water for 20 minutes while still vacuum-packed. The thawed tissue's surface was then rinsed in tap water after removing the vacuum seal. To maintain the integrity of the tissue samples, all scanning was done in a bath of degassed phosphate buffered saline (dPBS). To obtain this, a 10 times PBS concentrate (DNase, RNase and protease free filtered through a 0.2 μm filter; pH: 7.4 +/- 0.1 contains: 11.9 mM phosphates; 137 mM NaCl; 2.7 mM

KCl, (Fisher Scientific UK Ltd, Loughborough, UK) was diluted with distilled water at a volume ratio of 1:10 and boiled for 3 minutes to degas. It was then left to cool down at room temperature overnight.

For ease and repeatability of scanning, a common tray was used which was cleaned and re-prepared for each scanning session. A 2 cm thick piece of acoustic absorber (Aptflex F28, Precision Acoustics, Dorset, UK) was placed in the bottom of the tray to prevent artefacts from stray echoes and to assist tissue pinning. To offset the tissue from the acoustic absorber base and to provide a ground truth reference for the attenuation measurements, agar was added to the containers to a height of ~4 cm. Agar-Agar granular powder, (general purpose grade, Fisher Scientific UK Ltd.) was mixed with distilled water at a mass ratio of 1:99 and boiled for 10 minutes on a hotplate whilst stirring with a magnetic stirrer. After cooling for 15 minutes at room temperature, the agar was refrigerated for 2 hours, then allowed to settle at room temperature overnight.

For each sample, a 1 - 2 mm thick layer of acoustic coupling gel was applied to the agar surface prior to pinning the tissue on top. This was used to isolate the tissue signal from the agar substrate as well as to avoid air pockets. The tissue was then immersed in dPBS to a depth sufficient to allow the ultrasound transducer to be fully submerged while maintaining the tissue in the focal zone.

3.2. Microsphere Perfusion
To prepare the bowel tissue for perfusion with fluid and the microspheres, the whole tissue was pinned down and a suitable mesenteric vessel identified and cannulated. 1 - 2 ml of red-dyed fluid were then perfused to verify the cannulation. Upon successful cannulation, a perfusion pump (Braun Perfusor, Braun, Kronberg, DE) was used to perform the main perfusion. A commercial pump was used because its over-pressure alarm prevented pressures being attained which could damage the vasculature. Once perfused, the bowel samples were opened by cutting along the upper edge of where the mesentery inserted with a scalpel, exposing the inner luminal surface of the tissue for mechanical scanning.

Control samples were infused with 50 ml of dPBS at 200 ml/hr. Microsphere samples were infused with 30 ml of dPBS at 200 ml/hr, followed by infusion of 1μm diameter fluorescent microspheres (Polysciences Inc, PA, USA) diluted in 10 ml dPBS solution at variable dilution and perfusion rates to test the impact of dilution and diffusion rate on propagation within the tissues. A final 10 ml of dPBS was then used to flush the line and ensure the same total fluid infusion between control and microsphere samples.

Surface microspheres were flushed from the samples using dPBS and the penetration of the microspheres into the sample was analysed using a fluorescent lamp (Figure 1).

3.3. Oesophageal Samples
Similar to the bowel samples, the oesophageal samples were prepared for scanning by bisecting them along the long axis using a scalpel. As the attenuation measurements require the use of a reference echo which has passed through only the tissue of interest, in this case the upper two layers, consisting of the mucosa and submucosa, two oesophageal samples were prepared using the same method as that used for the whole samples, but were then separated along the fascial plane of the tissue via blunt dissection. The mucosa - submucosal combination was separated from the remaining layers, consisting of the muscularis propria and serosa and prepared for imaging the same way as the whole tissue.

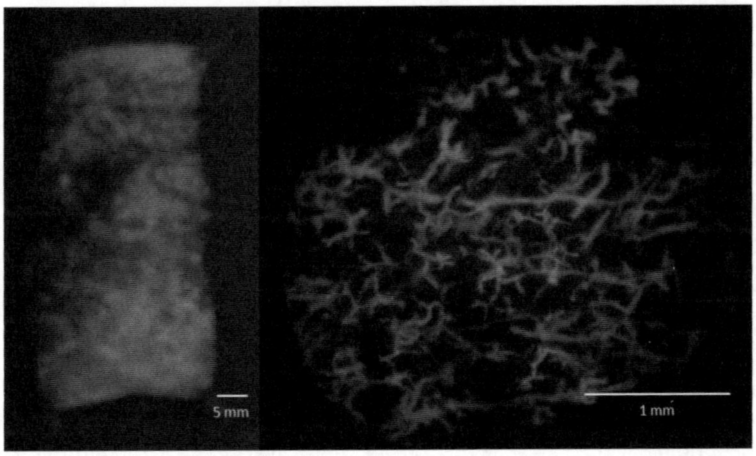

Figure 1 Microsphere-infused (1:10 volume ratio) *ex vivo* bowel tissue illuminated with a fluorescent light source. Profusion within the entire tissue sample can be seen on the left, while a 40 times magnification (right) shows that the microspheres have transited through the capillary network, but not infiltrated the interstitium or larger cellular network.

4. Scanning System

To ensure high-quality, repeatable microultrasound scans, an automated mechanical scanning system was used in all experiments to obtain the image data [14]. This system has a Labview-based (National Instruments, Austin, US) graphical user interface (GUI) used to control a pair of orthogonally-mounted linear motors supporting the scanning head of the transducer, as well as handling the triggering and data capture for the pulser/receiver unit (DPR500, JSR Ultrasonics, Pittsford, USA) and oscilloscope (MDO3014, Tektronix Inc, Beaverton, OR, USA) (Figure 2).

4.1. Scanning Software

The custom-programmed Labview GUI used to operate the microultrasound scanning system allows user control of the physical extents of each B-scan, the step size between A-scans, the time window used for capture for the RF data and any time offset. It also synchronises the triggering of the pulser/receiver unit, calibrates the digitising oscilloscope and displays and stores the data in general-use text files. The transducer used in all scanning reported here is a 48 MHz single-element transducer focused at 6 mm (AFMTH19, AFM Ltd, Birmingham, UK). The transducer was manufactured from a micromoulded PZT composite and at its focus has an axial resolution of 32 μm and a lateral resolution of 170 μm. Scans were aligned to place the centre of the tissue in the focal zone of the transducer.

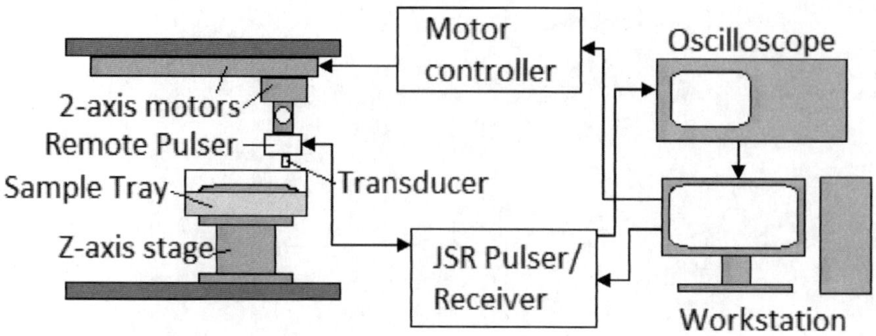

Figure 2 Mechanical scanning system used to collect the microultrasound scan data from the oesophageal and small bowel tissue samples. Communications are controlled by a custom Labview GUI on the central terminal and data is captured by the digitising oscilloscope before transfer to the terminal for display and storage.

4.2. Software Segmentation

For the acoustic property analysis techniques presented in this paper to be applicable to use *in vivo*, tissue identification and isolation must be performed digitally to ensure that the quantitative values obtained are not biased by heterogeneous tissue data. For this purpose, segmentation algorithms were developed in MATLAB (Mathworks, Natick, USA) and applied to the RF data obtained from the oesophageal tissue. The images were segmented with software masks, masking off tissue echoes below an adjustable threshold, then smoothing and hole-filling were performed to obtain a bulk zone containing the tissue of interest. Clinical judgement was used to determine the appropriate threshold for one tissue sample, then tested against the other five samples. The resulting assessment was that the mucosa/sub-mucosa had been successfully isolated from the lower layers based on a single, -26 dB threshold, referenced to the maximum echo amplitude in the image.

4.3. Quantitative Calculations

To derive the acoustic properties of the oesophageal tissue described in Section 2.2, full RF data scans were collected from the two mechanically separated tissue samples as well as a reference scan from a sample tray without a sample but with the same acoustic absorber/agar substrate and dPBS imaging medium. These data were then used to evaluate Eqn. 1 for each mechanical scan position and the mean and standard deviation were determined for each sample, as well as a weighted average for the overall sample set. A reference scan was obtained with the same agar substrate with a 0.5 mm thick quartz flat (Boston Piezo-Optics Inc., Bellingham, USA) on top to serve as a reference for the Z and BSC calculations.

5. Results

5.1. Qualitative Scans

As shown in Figure 1, while the microspheres were successfully diffused into the full breadth of the tissue sample, they were unable to extravasate the walls of the vascular system to properly simulate tissue inflammation. However, microultrasound scans performed before and after infusion, Fig. 3, show that strong μUS contrast is obtained from the microspheres and that their larger clusters can be identified and localised from qualitative imaging.

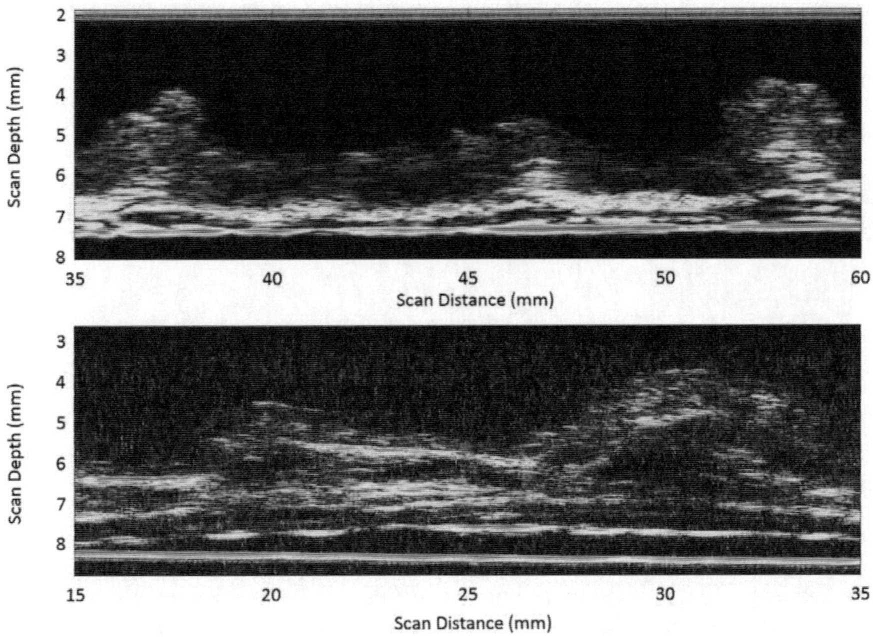

Figure 3 48 MHz μUS scans of the bowel tissue before (above) and after (below) microsphere infusion at 100 ml/hr with a 1:30 volume ratio. Microsphere concentrations can be particularly noted at the 30 mm position in the lower scan as well as by increased echo throughout the scan. Scan depth is measured with respect to the front face of the microultrasound transducer.

5.2. *Quantitative Measurements*

The mechanically separated oesophagus scans were used to calculate mucosal attenuation values, Fig. 4, and values of $\alpha = 1.92 \pm 0.77$ dB mm^{-1} and $\alpha = 1.78 \pm 0.66$ dB mm^{-1} were obtained as averages across each sample. $\alpha_{av} = 1.86 \pm 0.72$ dB mm^{-1}, averaged over all A-scans, was then used in the digitally separated samples.

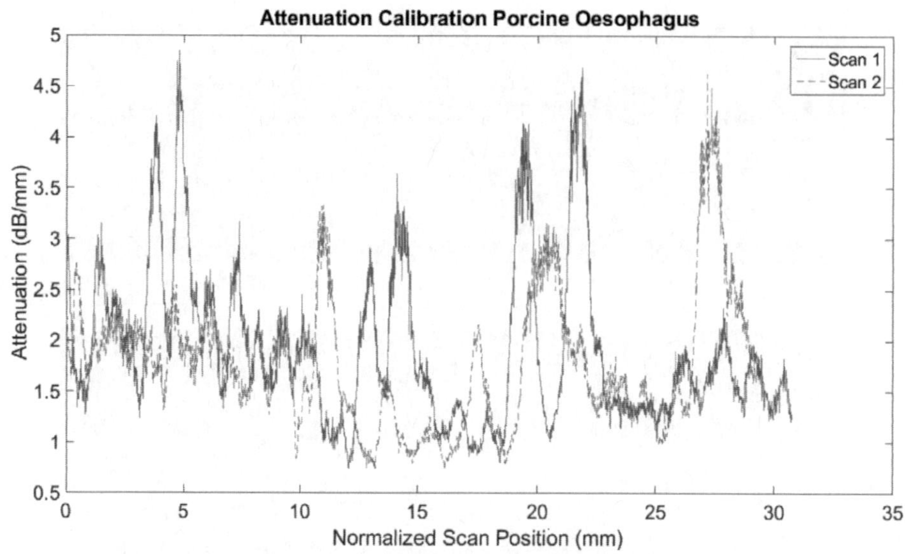

Figure 4. Attenuation as calculated from the mechanically separated oesophageal samples and agar reference substrate. Scan position is defined along the length of the tissue (oesophagus to stomach) referenced to the start of the physical scan. The two scans are for two distinct tissue samples.

The acoustic impedances and the backscatter coefficients of the six digitally segmented samples were computed and plotted for intra- and inter-sample comparison, Figs. 5 and 6. The overall mean and standard deviation values are shown in Table 1.

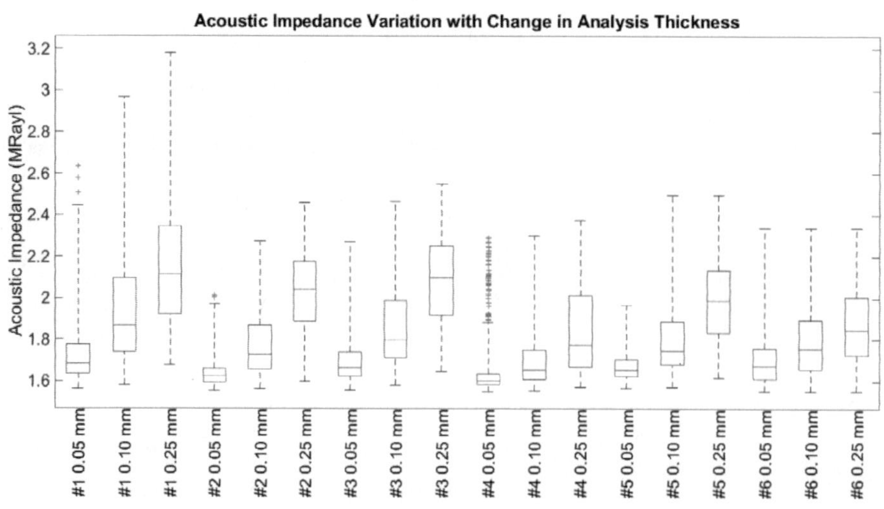

Figure 5. Acoustic Impedance of the front surface of the digitally segmented tissue samples as a function of front surface tissue thickness and sample number. Variation within each measurement is shown, with box plot depicting median, 25/75% boundary and maximum and minimum, with outliers beyond 5 times the box width marked with crosses.

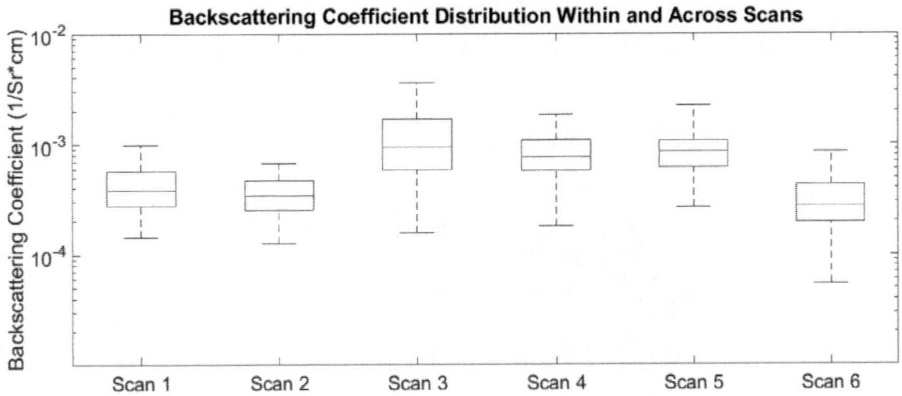

Figure 6 Backscattering coefficients for each digitally segmented tissue sample as a function of sample. Box plots indicate median, 25 and 75% extents with whiskers indicating maximum and minimum.

Table 1. Material Properties calculated for the digitally segmented porcine oesophageal samples based on the attenuation values calculated from the mechanically separated tissues.

Sample #	Z_{50} (MRayl)	Z_{100} (MRayl)	$Z2_{50}$ (MRayl)	BSC ($Sr^{-1}cm^{-1}$)
1	1.73 ± 0.15	1.95 ± 0.28	2.16 ± 0.29	$4.25 * 10^{-4} \pm 1.80 * 10^{-4}$
2	1.64 ± 0.06	1.76 ± 0.16	2.04 ± 0.18	$3.58 * 10^{-4} \pm 1.29 * 10^{-4}$
3	1.71 ± 0.12	1.86 ± 0.19	2.09 ± 0.21	$1.18 * 10^{-3} \pm 7.57 * 10^{-4}$
4	1.62 ± 0.08	1.70 ± 0.14	1.85 ± 0.21	$8.28 * 10^{-4} \pm 3.51 * 10^{-4}$
5	1.67 ± 0.07	1.80 ± 0.16	1.99 ± 0.19	$8.70 * 10^{-4} \pm 3.68 * 10^{-4}$
6	1.70 ± 0.11	1.79 ± 0.16	1.87 ± 0.17	$3.24 * 10^{-4} \pm 1.69 * 10^{-4}$

6. Discussion and Conclusions

Examination of the pre-infused microultrasound GI tissue scans show good layer differentiation. A clinician was able to identify the boundaries and thickness of the tissue layers from the B-scans, indicating that sufficient resolution was achieved for qualitative image analysis.

In the microsphere perfused scans, the microspheres could be distinguished visually, both optically and with the μUS scans, allowing verification of the scan data across the two imaging modalities. However, the lack of infiltration beyond the vascular network means that further work is required to fully mimic the behaviour of tissue inflammation. The microspheres showed good contrast in the μUS scans and the overall quality of the images validated the use of porcine tissue samples and the imaging modality for continued work in GI tissue characterisation.

The attenuation data collected from the two mechanically separated oesophageal tissue samples showed high intra-sample variation but good inter-sample agreement. This and the regularity of the variance seen in the plots suggests that high variance is normal across healthy tissue, possibly because of its convoluted surface topography. The inter-sample agreement, however, provides some confidence in the use of mean values for non-mechanically separated samples.

Overall, the acoustic impedance values calculated showed good agreement, with an increase in acoustic impedance with increasing tissue thickness. This is consistent with the known microstructure of the tissue, Fig. 7, with a highly irregular mucosal surface with deep crypts crowned with villi in the epithelium. The increase in acoustic impedance may be caused by the irregularity of the surface of the tissue columns, with more columns being included in the calculations for thicker segments. This rise in acoustic impedance with digital segmentation thickness could provide a standard against which diseased tissue could be assessed for breakdown in the regular structures caused by cellular transformation and architectural disruption.

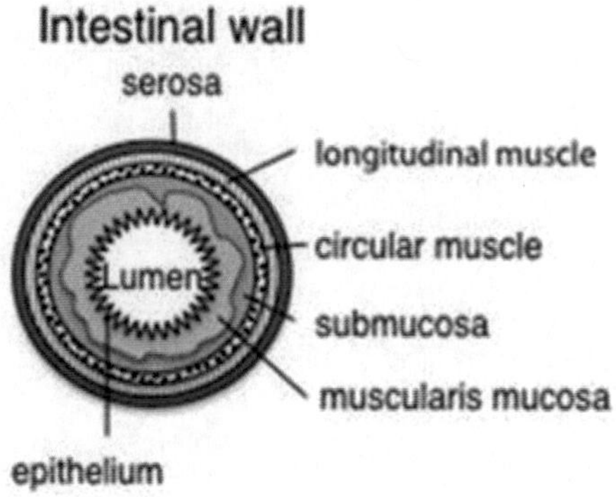

Figure 7. Anatomy of the tissue layers of the human gastrointestinal tract. All mechanical scans were performed with the epithelium facing the transducer.

The BSC values show larger variance than the acoustic impedance values, in agreement with the variability seen in the attenuation values and in the literature [15]. While studies of diseased tissue will be necessary to assess the change in mean value caused by disease progression, further steps can also be taken to reduce the large standard deviation seen in the measurements. The longitudinal plots of the attenuation values obtained from the mechanically separated samples shows relatively high variation across the tissue. This variation may contribute to the error in BSC calculations performed on the digitally segmented tissues as a mean value of attenuation was used in those measurements.

The studies presented in this work demonstrate the potential for the use of μUS imaging for GI tissue characterisation quantitatively as well as qualitatively, but challenges remain to maximise the potential. Furthermore this work illustrates the potential qualitative and quantitative μUS has in the diagnosis and management of GI disorders. The qualitative aspects of high resolution μUS would provide direct evidence of developing and evolving pathology. The quantitative facet has the potential to provide a two pronged advantage in terms of presenting data in an objective (metric'd) manner that that can be adapted into computer assisted diagnosis (CADx) thus aiding clinicians in reducing read times already associated with standard capsule endoscopy.

Acknowledgements
This work was funded by the Engineering and Physical Sciences Research Council (EPSRC) Sonopill grant (EP/K034537/1).

References

[1] U. C. Ghoshal, "Capsule Endoscopy: A New Era of Gastrointestinal Endoscopy," in *Endoscopy of GI Tract*, S. Amornyotin, Ed. InTech, 2013.
[2] G. Ciuti, A. Menciassi, and P. Dario, "Capsule Endoscopy: From Current Achievements to Open Challenges," *IEEE Rev. Biomed. Eng.*, vol. 4, pp. 59–72, 2011.
[3] B. F. Mustafa, M. Samaan, L. Langmead, and M. Khasraw, "Small bowel video capsule endoscopy: an overview," *Expert Rev. Gastroenterol. Hepatol.*, vol. 7, no. 4, pp. 323–329, May 2013.
[4] M. Rendi, "Crohn Disease Pathology: Overview, Epidemiology, Etiology," Apr. 2016.

[5] S. Ødegaard, L. B. Nesje, O. D. Lærum, and M. B. Kimmey, "High-frequency ultrasonographic imaging of the gastrointestinal wall," *Expert Rev. Med. Devices*, vol. 9, no. 3, pp. 263–273, May 2012.

[6] S. Sharma, "Micro-Ultrasound Imaging of Tissue Dysplasia," Doctoral Thesis, University of Dundee, 2015.

[7] R. K. Saha and M. C. Kolios, "Effects of cell spatial organization and size distribution on ultrasound backscattering," *IEEE Trans. Ultrason. Ferroelectr. Freq. Control*, vol. 58, no. 10, pp. 2118–2131, Oct. 2011.

[8] M. L. Oelze, J. F. Zachary, and W. D. O'Brian, Jr, "Characterization of tissue microstructure using ultrasonic backscatter: Theory and technique for optimization using a Gaussian form factor," *J. Acoust. Soc. Am.*, vol. 112, no. 3, pp. 1202–1211, Sep. 2002.

[9] E. J. Feleppa, T. Liu, A. Kalisz, M. C. Shao, N. Fleshner, V. Reuter, and W. R. Fair, "Ultrasonic spectral-parameter imaging of the prostate," *Int. J. Imaging Syst. Technol.*, vol. 8, no. 1, pp. 11–25, Jan. 1997.

[10] J. Jankowski, H. Barr, K. Wang, and B. Delaney, "Diagnosis and management of Barrett's oesophagus," *BMJ*, vol. 341, p. c4551, 2010.

[11] L. Schantz, K. Laber-Laird, S. Bingel, and M. Swindle, *Essentials of Experimental Surgery: Gastroenterology*. CRC Press, 1996.

[12] H. W. Siesler, "12.3.1 Influence of Fixation on Acoustic Parameters of Cells and Tissues," in *Biomedical Imaging: Principles and Applications*, John Wiley & Sons, Inc., 2012.

[13] F. S. Foster, M. Strban, and G. Austin, "The Ultrasound Macroscope: Initial Studies of Breast Tissue," *Ultrason. Imaging*, vol. 6, pp. 243–261, 1984.

[14] T. Anbarasan, C. Demore, H. Lay, M. Sunoqrot, R. Poltarjonoks, S. Cochran, and B. Cox, "Development of small bowel tissue phantom for microultrasound (uUS) investigation," in *Biosensor Technologies*, A. Rasooly and B. Pickril, Eds. Springer, 2016.

[15] C. M. Moran, N. L. Bush, and J. C. Bamber, "Ultrasonic Propagation Properties of Excised Human Skin," *Ultrasound Med. Biol.*, vol. 21, no. 9, pp. 1177–1190, 1995.

IOP Conf. Series: Journal of Physics: Conf. Series **797** (2017) 012004 doi:10.1088/1742-6596/797/1/012004

A model to predict the ultrasonic field radiated by magnetostrictive effects induced by EMAT in ferromagnetic parts

B Clausse[1], A Lhémery[1] and H Walaszek[2]
[1] CEA LIST, F-91191, Gif-sur-Yvette, France
[2] CETIM, F-60304, Senlis, France

E-mail: alain.lhemery@cea.fr

Abstract. An Electro-Magnetic Acoustic Transducer (EMAT) is a non-contact source used in Ultrasonic Testing (UT) which generates three types of dynamic excitations into a ferromagnetic part: Lorentz force, magnetisation force, and magnetostrictive effect. This latter excitation is a strain resulting from a magnetoelastic interaction between the external magnetic field and the mechanical part. Here, a tensor model is developed to transform this effect into an equivalent body force. It assumes weak magnetoelastic coupling and a dynamic magnetic field much smaller than the static one. This approach rigorously formulates the longitudinal Joule's magnetostriction, and makes it possible to deal with arbitrary material geometries and EMAT configurations. Transduction processes induced by an EMAT in ferromagnetic media are then modelled as equivalent body forces. But many models developed for efficiently predicting ultrasonic field radiation in solids assume source terms given as surface distributions of stress. To use these models, a mathematical method able to accurately transform these body forces into equivalent surface stresses has been developed. By combining these formalisms, the magnetostrictive strain is transformed into equivalent surface stresses, and the ultrasonic field radiated by magnetostrictive effects induced by an EMAT can be both accurately and efficiently predicted. Numerical examples are given for illustration.

1. Ultrasonic Testing using EMAT in ferromagnetic materials

In Ultrasonic Testing (UT), elastic waves are generated into the solid under examination by transducers. In most cases, piezoelectric transducers are used, operating from outside through a medium which mechanically couples them to the solid. When the use of such coupling is avoided or if piezoelectric transducers are not adapted to the need, the implementation of alternative non-contact solutions is required. For this, the piece under examination is excited by an external source of a physical field of non-mechanical nature (e.g. electromagnetic). The interaction of this field with the elastic part converts the non-mechanical energy involved into dynamic mechanical effects within the piece material, which in turn becomes a source of elastic waves. Electro-Magnetic Acoustic Transducers (EMAT) [1,2] are among the main non-contact probes currently used in UT. Since no coupling medium is needed, such sensors are practically suitable to operate in harsh conditions (high gradients of pressure, temperature). Another attractive aspect of EMAT consists in its ability to excite multiple modes with a high reproducibility, giving the opportunity to design various testing applications [3]. However, EMATs usually present low signal-to-noise ratios (compared to piezoelectric ones), and are strongly configuration and material dependent which brings some complexities to the modelling process. Generally, EMATs are made of one (or several) permanent magnet(s) and coil(s). The dynamic transduction processes they generate are produced by electromagnetic fields from the different constituting elements through a lift-off usually assumed to be made of air. The mechanical sources are generated in a thin layer below the solid free (of coupling medium) surface. The layer thickness depends on both characteristics of the excitation fields involved and on the ability of the solid medium to transform them into dynamic mechanical energy. The operating principle of an EMAT is shown in figure 1:

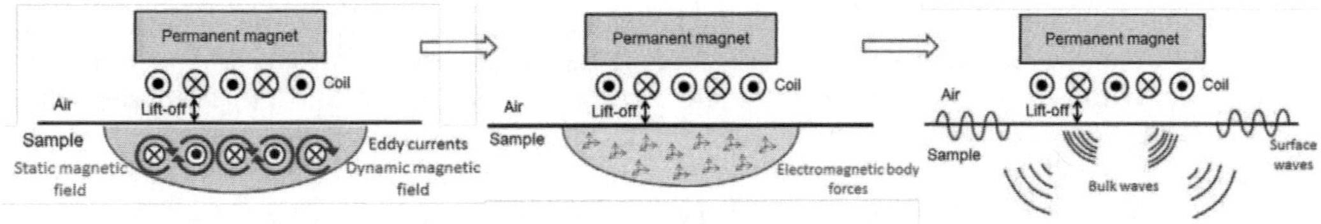

Figure 1. Operating principle of an ElectroMagnetic Acoustic Transducer (EMAT).

Three main electromagnetic dynamic excitations coexist in case of ultrasounds generation by EMAT into a ferromagnetic material. They are discussed in the following subsections.

1.1. Lorentz force
This fundamental phenomenon appears in any conductive material. It represents the force which acts on moving charges [4,5]. In EMAT applications, the magnetic Lorentz force stems from the interaction between the eddy currents density $\mathbf{J}$ $[\mathrm{A.m^{-2}}]$ and the magnetic induction field $\mathbf{B}$ [T] in the material, with a static component produced by the permanent magnet and a dynamic one generated by the coil. The Lorentz body force per unit volume $[\mathrm{N.m^{-3}}]$ is commonly expressed as:

$$\mathbf{f}^{\mathbf{L}} = \mathbf{J} \times \mathbf{B}. \quad (1)$$

1.2. Magnetisation force
This dynamic excitation is specific to ferromagnetic materials and can be understood as the force acting on atomic magnetic moments inside the material in the presence of a spatially varying magnetic field. Indeed, ferromagnetic materials are divided up into magnetic domains, namely the Weiss domains separated by Bloch interfaces, within which the orientation of all the particle magnetic moments are aligned along the same direction [6]. The application of an external magnetic field on a ferromagnetic material reorganizes the orientation of these atomic magnetic moments in order to align them along the external magnetic field direction, as illustrated below in figure 2:

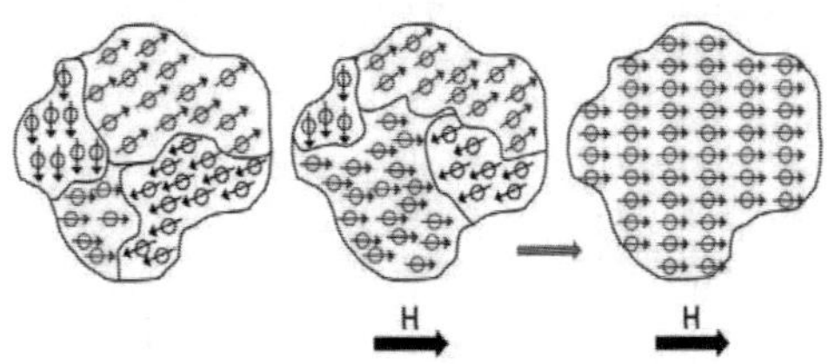

Figure 2. Weiss domains and atomic magnetic moment reorientation under external magnetisation.

The magnetisation process is then the force responsible for the reorientation of these atomic magnetic moments inside the sample, and corresponds finally to a macroscopic evidence of microscopic phenomena. The expression of the magnetisation body force per unit volume $[\mathrm{N.m^{-3}}]$ takes several forms in the literature which can lead to theoretical confusion, principally due to the choice of the magnetic fields involved in this phenomenon. It is not the purpose of this work to properly address the question of what formulation is the most appropriate. Denoting the magnetisation field by $\mathbf{M}$ $[\mathrm{A.m^{-1}}]$ and the magnetic field by $\mathbf{H}$ $[\mathrm{A.m^{-1}}]$, the expression of the magnetisation body force adopted here (taken from [4,5]) is given by the relation:

$$\mathbf{f^M} = \mu_0(\mathbf{M}.\nabla)\mathbf{H}, \tag{2}$$

where μ_0 [$\mathrm{H.m^{-1}}$] is the magnetic permeability in vacuum.

1.3. Magnetostrictive effects

This phenomenon is also specific to ferromagnetic materials. It is related to magnetoelastic interactions at the material microscopic scale. Precisely, the 'magnetostriction' term stands for all the elastic variations of a magnetic material under external magnetic fields. Several types of magnetostrictive effects (direct or inverse) are referenced in the literature [5,6]. Here, the discussion is focused on the longitudinal Joule's magnetostriction, cited as the main impacting effects in our applications area [6,7]. These magnetostrictive effects result from the displacement of Bloch interfaces within the ferromagnetic part under external magnetisation, which creates a strain at the macroscopic scale measured along the magnetisation direction. Like the magnetisation force, the magnetostrictive effects depict a macroscopic manifestation of microscopic phenomena, which can be illustrated by figure 3:

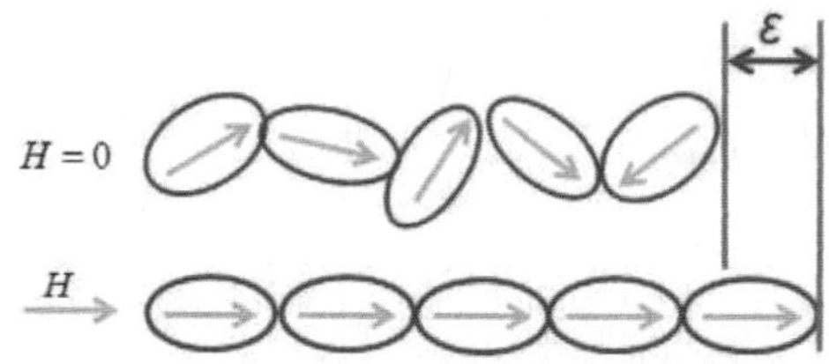

Figure 3. Magnetostrictive strain measured along the external magnetic field.

As mentioned, this phenomenon reduces at the macroscopic scale to a dimensional change (a strain). For the sake of an efficient semi-analytical model of ultrasound generation by EMAT, a model allowing the transformation of this magnetostrictive deformation into an equivalent body force of magnetostriction is needed. This is the purpose of the next section.

2. Modelling the equivalent body force of magnetostriction under a strong static magnetic field

The aim here is to establish an equivalent body force model which faithfully represents the direct longitudinal Joule's magnetostriction phenomenon previously described, that is to say, the macroscopic material deformation under an external magnetisation varying both with time and space. We assume first that these effects are magnetically isotropic at macroscopic scale and create isovolume dimensional change. These assumptions are commonly encountered in the literature [2,6,8], but an extension of the model could easily be done if needed to avoid making them. A weak magnetoelastic coupling is also assumed inside the ferromagnetic part. This could be justified by considering that the involved electromagnetic interaction volume is only restricted to a thin surface layer of the material which is limited by the electromagnetic skin depth:

$$\delta = \sqrt{\frac{1}{\pi f \mu \sigma}}, \tag{3}$$

where f is the inspection frequency, μ is the magnetic permeability, and σ the electric conductivity of the material being tested. Since the electromagnetic skin depth δ is much smaller than the wavelength λ of the acoustic wave produced in most EMAT applications ($\delta \ll \lambda$), the elastic waves which propagate below the surface into the part are assumed not to appreciably alter the electromagnetic fields confined in this close electromagnetic interaction vicinity of the part surface.

Under such assumptions, the electromagnetic quantities are known by solving the classical set of Maxwell's equations inside the concerned material volume (in a quasi-static approximation considering the inspection frequencies generally below than 10 MHz in our application cases). The problematic addressed here is then to rigorously formulate the magnetostrictive transduction process starting from the knowledge of these electromagnetic fields. Inspired by [9], the idea in the present work is to model the three-dimensional magnetostrictive strain ε_{ij}^{MS} from the magnetic field $\mathbf{H}$ as:

$$\varepsilon_{ij}^{MS}(\mathbf{H}) = \lambda(||\mathbf{H}||)S_{ij}(\mathbf{H}). \tag{4}$$

This tensorial formulation involves two main terms which respect all the assumptions being made:

- The elementary magnetostrictive strain λ is measured in the direction of magnetisation and is only dependent on the studied material and the magnetic field intensity $||\mathbf{H}||$. This parameter is assumed to be a scalar in order to fulfil the macroscopic magnetic isotropy hypothesis, and can be obtained for example by interpolation of experimental data, as illustrated in figure 4:

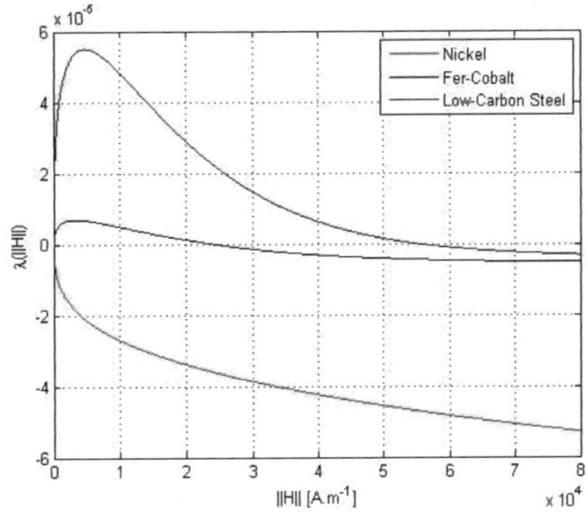

Figure 4. Dependency of the elementary magnetostrictive strain with the magnetic field intensity.

The formulation can also be easily extended to the case of magnetically anisotropic materials by considering $\boldsymbol{\lambda}$ as a second order tensor, written for example in the basis of easy magnetisation axes [6,10].

- The magnetic direction tensor $\mathbf{S}$ is a second order symmetric tensor only depending on the direction of the total magnetic field $\mathbf{H}$, and is expressed as:

$$\mathbf{S}(\mathbf{H}) = \frac{1}{2||\mathbf{H}||^2}(3\mathbf{H}\otimes\mathbf{H} - ||\mathbf{H}||^2\mathbf{I}), \tag{5}$$

where the symbol $\otimes$ denotes the tensor product and $\mathbf{I}$ is the second order identity tensor. The tensor $\mathbf{S}$ is defined irrespective of the coordinate system used and helps modelling an isovolume dimensional change by verifying a null trace. It formulates this assumption by catching the principal direction of the total magnetic field and by assigning half the opposite value in the two other perpendicular directions. For example, in case of a total magnetic field $\mathbf{H} = H_3\mathbf{e}_3$, this tensor takes the following form:

$$\mathbf{S} = \begin{pmatrix} -0.5 & 0 & 0 \\ 0 & -0.5 & 0 \\ 0 & 0 & 1 \end{pmatrix}. \tag{6}$$

Based on this representation of the magnetostrictive strain, the next fundamental mathematical step relies on the fact that the total magnetic field is a superposition of a static part (induced by the permanent magnet(s)) and of a dynamic part (induced by the coil(s)). In most EMAT applications, the intensity of the current flowing in the coil is small enough to ensure that the dynamic part of the magnetic field is much smaller than the static bias magnetic field. So, one has:

$$\mathbf{H} = \mathbf{H_S} + \mathbf{H_D}, \text{ with: } \mathbf{H_D} \ll \mathbf{H_S}. \tag{7}$$

Considering the magnetostrictive strain ε_{ij}^{MS} as a function of the magnetic field $\mathbf{H}$, its first order Taylor expansion around the static magnetic field $\mathbf{H_S}$ position leads to:

$$\varepsilon_{ij}^{MS}(\mathbf{H} = \mathbf{H_S} + \mathbf{H_D}) = \varepsilon_{ij}^{MS}(\mathbf{H_S}) + \frac{\partial \varepsilon_{ij}^{MS}}{\partial H_k}(\mathbf{H_S})H_{Dk} + o(\|\mathbf{H_D}\|^2). \tag{8}$$

This simple development is particularly suitable since it provides a rigorous mathematical formulation which enables a relevant physical interpretation of the magnetostrictive phenomenon under the assumption of a strong static magnetic field. Indeed, the first term $\varepsilon_{ij}^{MS}(\mathbf{H_S}) = \lambda(\|\mathbf{H_S}\|)S_{ij}(\mathbf{H_S})$ depicts the time-independent longitudinal magnetostrictive strain which appears along the static field, and which can be seen as a spontaneous initial state of the ferromagnetic material under the external static magnetic field. The second term is the time-dependent leading term which gives rise to the ultrasonic elastic waves into the sample. It involves the third order piezomagnetic strain tensor d_{ijk}^{MS} evaluated at the static magnetic position and defined by the partial derivative of the magnetostrictive strain over the total magnetic field. This tensor is widely used in the literature when dealing with the linearized magnetoelastic coupled equations (see for example [2,11]). Taking benefits from the analytical tensorial expression of $\mathbf{S}$, one can derive an analytical expression for the piezomagnetic strain tensor d_{ijk}^{MS} by means of some tensorial and differential analyses:

$$\mathbf{d^{MS}}(\mathbf{H}) = \frac{\partial \boldsymbol{\varepsilon}^{\mathbf{MS}}}{\partial \mathbf{H}} = \frac{\varphi(\|\mathbf{H}\|)}{\|\mathbf{H}\|}\mathbf{S}\otimes\mathbf{H} + \frac{3\lambda(\|\mathbf{H}\|)}{\|\mathbf{H}\|^2}\left[\frac{\mathbf{H}\underline{\otimes}\mathbf{I}}{2} - \frac{\mathbf{H}\otimes\mathbf{H}\otimes\mathbf{H}}{\|\mathbf{H}\|^2}\right], \text{with: } \varphi(\|\mathbf{H}\|) = \lambda'(\|\mathbf{H}\|). \tag{9}$$

In the previous expression, the operator $\underline{\otimes}$ between a vector $\mathbf{v}$ and a second order tensor $\mathbf{T}$ is given in Voigt notation by: $[\mathbf{v}\underline{\otimes}\mathbf{T}]_{ijk} = v_i T_{jk} + v_j T_{ik}$. This analytical tensorial expression is interesting since it models a general way to study the dynamic component of the magnetostrictive strain irrespective of the EMAT configuration and the geometry of the ferromagnetic part, which is quite missing in the literature. By using the elastic stiffness tensor $\mathbb{C}$, the piezomagnetic stresses can be written as:

$$\boldsymbol{\sigma}^{\mathbf{MS}} = -\mathbb{C} : [\boldsymbol{\varepsilon}^{\mathbf{MS}}(\mathbf{H_S}) + \mathbf{d^{MS}}(\mathbf{H_S}).\mathbf{H_D}], \tag{10}$$

where the symbol : denotes the tensorial double contraction product. This expression must be completed by taking into account free surface boundary condition in order to reproduce the physical context of the study (for example discussed in [1]). Denoting by $\partial\Omega$ the part surface and by $\mathbf{n}$ the outgoing normal vector to $\partial\Omega$, this condition is given by:

$$\boldsymbol{\sigma}^{\mathbf{MS}}.\mathbf{n} = \mathbf{0} \quad \text{on } \partial\Omega. \tag{11}$$

Finally, by taking the divergence of these piezomagnetic stresses, the modelling of equivalent body force of magnetostriction is achieved:

$$\mathbf{f^{MS}} = \boldsymbol{\nabla}.\boldsymbol{\sigma}^{\mathbf{MS}}. \tag{12}$$

The useful aspect of the mathematical approach in the present paper comes from its ability to deal with arbitrary coordinate system, allowing us to derive such a magnetostrictive equivalent body force for ferromagnetic parts of complex shape and irrespective of EMAT configuration. Of course, literature results for simpler cases can be straightforwardly derived from this model. For example, when an elastic homogeneous isotropic planar material in the x-y plane is considered with a static magnetic field along the depth of the material ($\mathbf{H_S} = H_s\mathbf{e_z}$), the formulation presented here reduces to the academic expression of the magnetostrictive body force given by [1,2]:

$$\mathbf{f^{MS}} = \begin{pmatrix} \mu_L\varphi(H_S)\frac{\partial H_{Dz}}{\partial x} - \frac{3\mu_L\lambda(H_S)}{H_S}\frac{\partial H_{Dx}}{\partial z} \\ \mu_L\varphi(H_S)\frac{\partial H_{Dz}}{\partial y} - \frac{3\mu_L\lambda(H_S)}{H_S}\frac{\partial H_{Dy}}{\partial z} \\ -2\mu_L\varphi(H_S)\frac{\partial H_{Dz}}{\partial z} - \frac{3\mu_L\lambda(H_S)}{H_S}\left[\frac{\partial H_{Dx}}{\partial x} + \frac{\partial H_{Dy}}{\partial y}\right] \end{pmatrix}, \tag{13}$$

where μ_L is the second Lamé's parameter (shear modulus).

By using the presented model of magnetostrictive equivalent body forces, the next section aims at giving a suitable mathematical process enabling to accurately and efficiently predict the ultrasonic field radiated by the dynamic magnetostrictive sources generated by EMAT.

3. Transformation into magnetostrictive equivalent surface stresses

3.1. General concept of mathematical transformation of body forces into equivalent surface stresses

In ultrasonic nondestructive evaluation (NDE), accurate knowledge of the ultrasonic field radiated by the source is of paramount importance as soon as quantitative results are expected, that is to say, if the method of examination is designed to go as far as defect characterization and sizing. For this reason, in the vast literature, many models have been developed to compute field radiation by ultrasonic sources. Among them, many models were developed to predict fields radiated by mechanical sources operating at the surface of the solid, as those generated by piezoelectric transducers in direct contact. The calculation of the elastic wave field radiated by a point-source of normal or tangential stress in an elastic half-space, that is to say, the Green's function of the so-called Lamb's problem, is the subject of hundreds of papers in the literature, including exact, approximate or numerical solutions and dealing with a variety of elastic media, after Lamb's pioneering contribution [12]. For finite-size sources, a simple surface and time convolution of the time-dependent surface stress distribution with one such solution for a point source leads to predict source diffraction effects [13]. It is our aim to use this kind of time and surface convolution models to predict the field radiated by non-contact sources. But since at this stage, all the dynamic excitations induced by EMAT in ferromagnetic materials are expressed as a volume distribution of electromagnetic body forces, a mathematical method is needed to accurately transform these body forces into equivalent surface stresses distribution, that is to say, which radiate the same ultrasonic field into the mechanical part.

Let us consider an open set Ω embodying the elastic domain of the part under test. We want to study the representation of particle displacement field $\mathbf{u}(\mathbf{x})$ radiated at an observation point $\mathbf{x} \in \Omega$ by a distribution of time-harmonic body force $\mathbf{f}(\mathbf{x_0})e^{i\omega t}$, where ω denotes the angular frequency and $\mathbf{x_0} \in \Omega$ denotes the source points. For clarity, the factor $e^{i\omega t}$ will be omitted in what follows. The exact way for determining the particle displacement field consists in a spatial volume convolution integral of body forces $\mathbf{f}(\mathbf{x_0})$ with the Green's tensor $\mathbf{G}(\mathbf{x},\mathbf{x_0})$ for elastic waves in Ω. So the idea of this section is to determine the most accurate surface distribution of equivalent stresses $\widetilde{\boldsymbol{\sigma}}^{\mathbf{f}}$ of the body force $\mathbf{f}$, which approximates the most faithfully the particle displacement $\mathbf{u}$ by a surface convolution with the same previous Green's tensor taken at the part surface. The approach is summed up by the relation:

$$\mathbf{u}(\mathbf{x}) = \iiint_{\mathbf{x_0}\in\Omega} \mathbf{G}(\mathbf{x},\mathbf{x_0}).\mathbf{f}(\mathbf{x_0})d\Omega \approx \iint_{\mathbf{X_0}\in\partial\Omega} \mathbf{G}(\mathbf{x},\mathbf{X_0}).\widetilde{\boldsymbol{\sigma}}^{\mathbf{f}}(\mathbf{X_0})d\Gamma. \tag{14}$$

The usual method for deriving equivalent surface stresses consists in a simple integration of the body forces over the depth (depicted by a variable w), which corresponds to the zeroth order force moment:

$$\widetilde{\boldsymbol{\sigma}}^{\mathbf{f}} = \mathbf{M_f^{(0)}} = \int_w \mathbf{f}(\mathbf{x_0})K(w)dw, \tag{15}$$

where K(w) denotes the local curvatures at a depth w of the part surface. But Thompson [14] pointed out in an ideal academic case (illustrated below in subsection 3.2.) that such a simple integration over the material depth predicts wrong stresses in case of magnetostrictive effects, mostly due to the free surface boundary condition previously discussed. This is the reason why Thompson [15], and more recently Rouge *et al.* [16], proposed a new model of equivalent surface stresses in elastic isotropic planar cases. It relies on a second order Taylor expansion in the direction normal to the part surface of the Green's integral formulation of the elastic wave equation and takes into account the moments of a body force distribution up to the second order, the p-th order force moment being denoted by $\mathbf{M}_\mathbf{f}^{(p)}$. In order to make this mathematical transformation applicable even broadly and so, consistently with the general magnetostrictive body force model presented in section 2, this approach has been recently extended to surfaces of complex shape and anisotropic elastic materials by means of tensorial and differential analyses [17]. The model takes into account the angular inspection frequency ω, the mass density ρ, the elastic stiffness tensor $\mathbb{C}$, the outgoing normal vector to the studied surface **n**, and can be reduced for the sake of clarity to the relation:

$$\tilde{\boldsymbol{\sigma}}^\mathbf{f} = \tilde{\boldsymbol{\sigma}}^\mathbf{f}\left(\mathbf{M}_\mathbf{f}^{(p)}, \omega^2, \rho, \mathbb{C}, \mathbf{n}\right), \quad p = 0,\ 1,\ 2. \tag{16}$$

Readers are referred to [17] for the detailed expression of this transformation. This mathematical model has also been numerically validated on bi-dimensional cases in order to demonstrate its accuracy gained in the particle displacement representation by a surface convolution. To maintain the process generality, this transformation method has been derived irrespective of nature of body forces considered. The only hypothesis is that these body forces satisfy a penetration depth much smaller than the acoustic wavelength λ, which is of course verified in EMAT applications considering that $\delta \ll \lambda$. The following subsections give 2D and 3D implementation examples of the magnetostrictive equivalent surface stresses distributions computed both by the usual method of simple integration (referenced as 'Order 0 method') and by the presented transformation model (referenced as 'Order 2 method').

3.2. *2D application example of the model in ideal EMAT configuration*

We present here the ideal and analytical 2D case historically studied by Thompson [14], which gave him the idea of a model based on a second order expansion of the equivalent surface stresses. It consists of a meander line coil carrying a current $Ie^{i\omega t}$ at the desired ultrasonic frequency ω and positioned below a permanent magnet delivering a static uniform magnetic field $\mathbf{H_S} = H_S\mathbf{e_x}$. Dynamic eddy currents and magnetic fields are set up at the surface of the ferromagnetic material (of relative magnetic permeability μ_R). Besides the uniform character of the static field, this case is ideal because it further assumes an infinite y extent and a dynamic magnetic field $\mathbf{H_D}$ only along the x direction (parallel to the static magnetic field). The EMAT configuration under study is illustrated in figure 5:

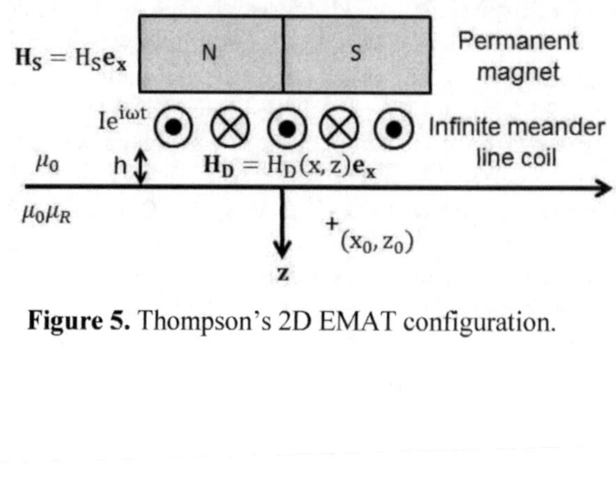

Figure 5. Thompson's 2D EMAT configuration.

As shown by Thompson [15], when this transducer has an infinite y extent and both static and dynamic fields in the x direction, it produces a force distribution on a ferromagnetic material with a component in the z-direction of the form:

$$f_z^{MS}(x_0, z_0) = \mu_L \varphi(H_S)[ik - \delta_D(z_0)]H_{Dx}(x_0)e^{-ikz_0}e^{i\omega t}, \tag{17}$$

where $k = (1 - i)/\delta$, (x_0, z_0) are the coordinates of a source point below the surface, and δ_D denotes a Dirac delta function at the part surface. In this ideal planar 2D case, it is worth noting that the model of equivalent magnetostrictive body force presented in section 2 reduces well to the magnetostrictive force given by Thompson, excepted for the Dirac delta function term. Indeed, this term is of non-physical nature and only artificially added at the end of the force derivation process by Thompson in order to satisfy the otherwise violated stress-free boundary condition. As mentioned above, the model presented here also takes into account a free surface boundary condition, but in the very definition of the piezomagnetic stresses, and so, without adding any non-physical term in the magnetostrictive force definition. Then, the zeroth and second order moments of the magnetostrictive normal body force can be calculated as follows:

$$M_{MSz}^{(0)}(x_0) = \int_{z_0=0}^{+\infty} f_z^{MS}(x_0, z_0)dz_0 = 0 \tag{18a}$$

$$M_{MSz}^{(2)}(x_0) = \int_{z_0=0}^{+\infty} z_0^2 f_z^{MS}(x_0, z_0)dz_0 = -\frac{2\mu_L\varphi(H_S)}{k^2}H_{Dx}(x_0)e^{i\omega t} \tag{18b}$$

Therefore, while the zeroth order moment of the magnetostrictive normal force vanishes, its second order moment is finite and non-zero. Thompson [15] made two independent observations to explain the inconsistency of predicting null surface stresses of magnetostriction in this case, which inspired him the idea of developing the equivalent surface stresses model up to a second order. The model was historically only designed for an elastic isotropic planar half-space, which has been corrected by Rouge *et al.* [16], and recently extended to media of complex shape and numerically validated [17]. In this special analytic configuration, the whole second order expansion of equivalent surface stresses model reduces to:

$$\sigma_z^{fMS}(x_0) = M_{MSz}^{(0)}(x_0) - \frac{\rho\omega^2}{2(\lambda_L+2\mu_L)}M_{MSz}^{(2)}(x_0) = \frac{\rho\omega^2\mu_L\varphi(H_S)}{(\lambda_L+2\mu_L)k^2}H_{Dx}(x_0)e^{i\omega t} \tag{19}$$

Besides the reasons given by Thompson in [15], this result has also been confirmed in [11] by considering the magnetostrictive interaction with ferromagnetic part as a pure inertial phenomenon, result that was numerically validated using finite-elements method. So, even if these equivalent surface stresses (being proportional to the ratio δ^2/λ^2) are weak in this ideal case, this study gives credit to the whole methodology since the equivalent surface stresses prediction by a second order expansion is non-zero. Moreover, the main advantages of the two mathematical models presented here (meaning the equivalent magnetostrictive body force and its equivalent surface stresses expanded to a second order) rely on their ability to be applicable in three-dimensional 'non-ideal' cases (that is to say without considering any infinite dimension extent) and irrespective of the EMAT configuration. The models allow therefore studying the surface variations of the electromagnetic fields produced by more complex configurations of EMAT with finite-size aperture over the excited surface. Results of such three-dimensional computations are given in the following subsection.

3.3. 3D computations of the equivalent surface stresses of magnetostriction

As mentioned above, we show here some computation results for three-dimensional configurations of EMAT with a finite-size aperture over the excited surface of a ferromagnetic material. Results in this subsection - meaning the computation of the EMAT configuration, the definition of the material elastic and electromagnetic properties, the calculation of the electromagnetic fields induced, the magnetostrictive body force and finally its equivalent surface stresses - were obtained by implementing the mathematical models presented in sections 2 and 3 in the Eddy Current Testing (ECT) modulus of the CIVA software platform [18] developed at CEA List and dedicated to simulation of various non-destructive evaluation techniques.

The assumptions made on the magnetic behaviour of the tested material need to be clearly described here. We first assume that the material does not exhibit an intrinsic permanent magnetisation; the magnetisation is thus purely induced by the magnetic excitation field. The constitutive relations of the ferromagnetic material are given by:

$$\mathbf{M} = \boldsymbol{\chi}(\mathbf{H})\,.\,\mathbf{H} \quad \text{and} \quad \mathbf{B} = \mu_0(\mathbf{H}+\mathbf{M}) = \boldsymbol{\mu}(\mathbf{H})\,.\,\mathbf{H},$$

where $\boldsymbol{\chi}$ is the material magnetic susceptibility, and $\boldsymbol{\mu} = \mu_0(\mathbf{1}+\boldsymbol{\chi}) = \mu_0\boldsymbol{\mu}_\mathbf{R}$ is the material magnetic permeability. Numerous theoretical models and experimental analyses are devoted to characterize these two key magnetic parameters, taking into account magnetic anisotropy, non-linear hysteretic behaviour, their dependencies on frequency, on temperature, or on applied stress. In order to make simpler our analysis on magnetostrictive effects, anhysteretic behaviour and magnetic isotropy are assumed in the present work and μ is modelled by a constant scalar parameter. This ideal case does not faithfully represent all possible experimental cases of magnetic behaviour; hysteretic behaviour and the dependency on frequency or on applied stress have a potential impact on the ultrasound generation processes involved in EMAT applications. Nevertheless, since a frequency domain simulation is performed with a weak coil injection current (below 1A) which presents a simple sinusoidal dependency with a frequency below 10 MHz, such assumptions constitute a good starting approximation to model soft ferromagnetic material behaviour focusing on the magnetostrictive model presented here.

The simulation case under study in the present paper is a 3D EMAT composed of a cubic permanent magnet with an induction intensity of 1 T positioned over a rectangular spiral coil with an excitation current intensity of 1 mA at an excitation frequency $f = 5$ MHz. The studied ferromagnetic material is a planar low-carbon steel with a relative magnetic permeability $\mu_R = 2$ and an electric conductivity $\sigma = 2\ \mathrm{MS.m^{-1}}$. It is further assumed elastically isotropic with the elastic quantities: $\rho = 7.8\ \mathrm{g.cm^{-3}}$, $c_P = 5900\ \mathrm{m.s^{-1}}$, and $c_S = 3230\ \mathrm{m.s^{-1}}$. The lift-off (in air) distance between the coil and the material is fixed at 0.2mm. An illustration of the studied EMAT configuration is given in figure 6:

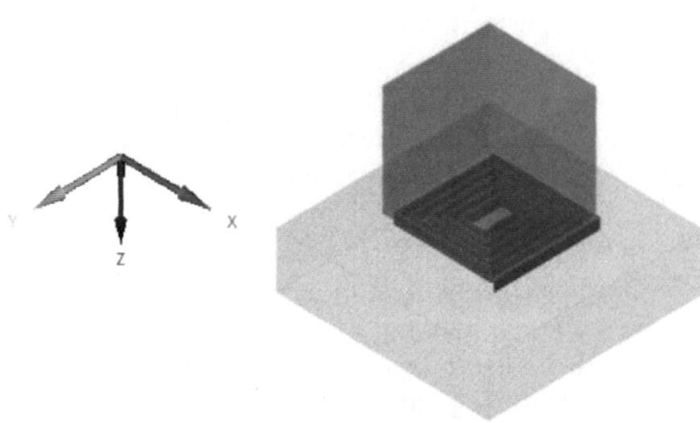

Figure 6. 3D EMAT composed of a cubic permanent magnet and a rectangular spiral coil.

Figure 7 displays the numerical simulation results using either a zeroth or a second order expansion of the equivalent surface stresses of the magnetostrictive equivalent body force induced by the EMAT in the detailed configuration, with a static magnetic induction $\mathbf{B_S}$ respectively along the x-direction (top) and along the z-direction (bottom).

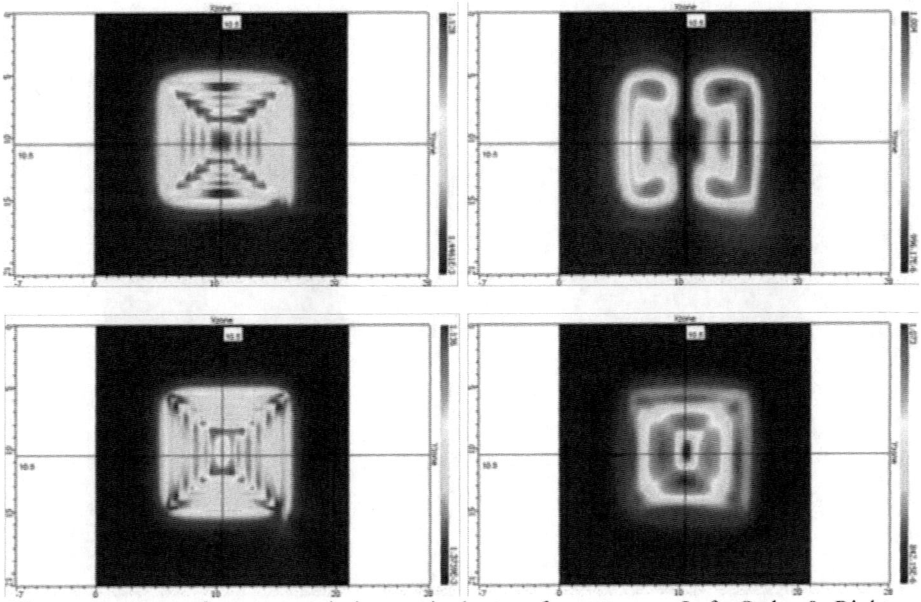

Figure 7. Norm of magnetostrictive equivalent surface stresses. Left: Order 0; Right: Order 2. Top: $\mathbf{B_S}$ along x. Bottom: $\mathbf{B_S}$ along z.

As we can observe on the norm of the equivalent surface stresses of magnetostriction, the zeroth order expansion (meaning the zeroth order force moment) of equivalent surface stresses does not vanish anymore as in the above 2D case, since the x and y components of the electromagnetic fields and of the surface stresses coexist. We can conclude from these figures that the magnetostrictive phenomenon is much more precisely predicted using a second order expansion model of equivalent surface stresses, since the results between the two models are quite radically different. An explanation can be found in [11] in considering the magnetostriction phenomenon as a pure inertial force. In fact, the second order expansion model of the equivalent surface stresses accounts for the inertial effects brought by the elastic wave equation; this is actually why the model is sensitive to the square of the angular frequency ω^2. The discussion confirms that magnetostriction effects are mostly due to inertial phenomenon, which can be modelled by the equivalent surface stress model detailed in section 3.

The next section uses these magnetostrictive equivalent surface stresses as source terms to predict the longitudinal (L) and transversal (T) elastic waves generated by this configuration of EMAT into the mechanical part being tested.

4. Simulation of the ultrasonic field radiated by magnetostrictive effects induced by EMAT

Finally, the modelling work done in the two previous sections allows taking benefits from well-established semi-analytical models of ultrasonic field radiation to predict the elastic wave field radiated by magnetostrictive effects induced by an EMAT into a ferromagnetic part. The setup here is the one described in the above section, using the detailed EMAT with a static magnetic induction $\mathbf{B_S}$ along the z-direction. Figures 8-9 depict respectively the longitudinal and transversal waves radiated into planar elastically isotropic low-carbon steel at a frequency $f = 5$ MHz. Results are given using either a zeroth or a second order expansion of the equivalent surface stresses of the magnetostrictive equivalent body force, and the relative differences ϵ_r (in percent) between the two transformation methods is represented according to the relation:

$$\epsilon_r = \left|\frac{u^{order2} - u^{ordero}}{\max(u^{order2}, u^{ordero})}\right|. \tag{20}$$

Maximum ultrasonic amplitude of the particle displacement modulus (in arbitrary units) radiated into the material is shown for a set of calculation points in a plane perpendicular to the piece surface (first row) and in a plane parallel to the surface at a depth of 50 mm (second row), for longitudinal waves (figure 8) and for transversal wave (figure 9).

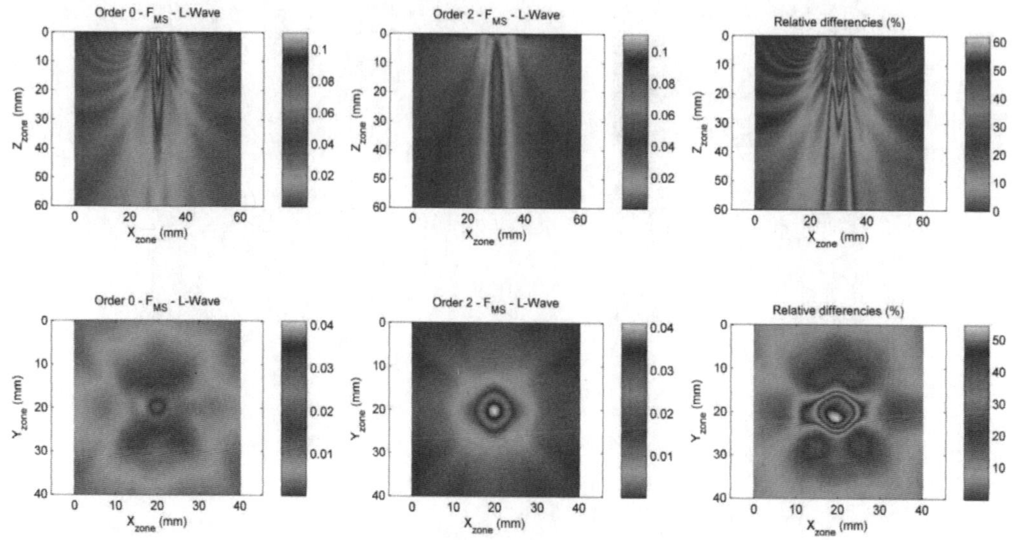

Figure 8. Map of the maximum L-wave amplitude induced by magnetostrictive effects. 1st row: in a plane *xz*, 2nd row; in a plane *xy* at a depth of 50 mm. 1st column: computed at the 0-th order, 2nd column: computed at the 2nd order, 3rd column: relative difference.

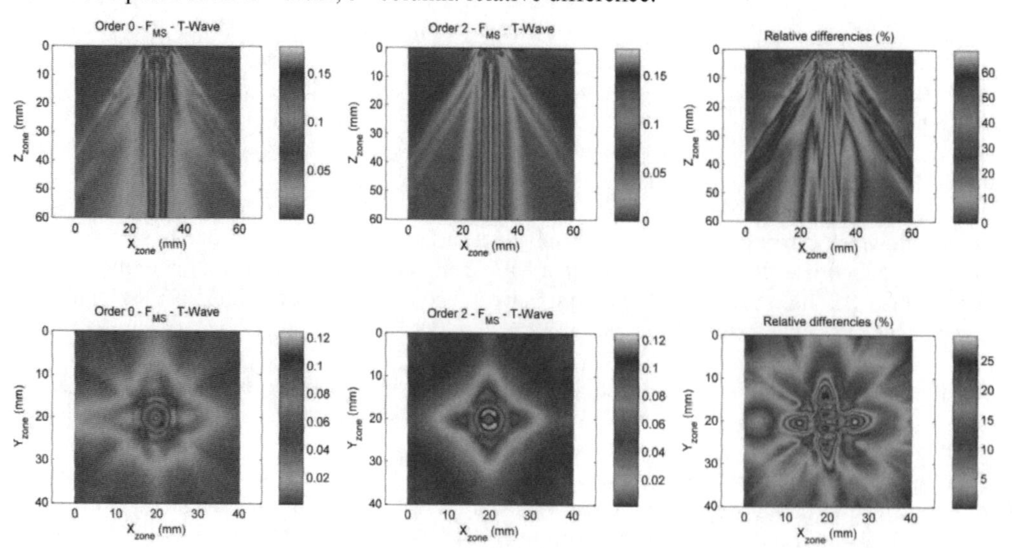

Figure 9. Same as figure 8, but T-wave radiation.

As expected in taking a look at the magnetostrictive equivalent surface stresses with either a zeroth or a second order expansion, ultrasonic fields radiated by these two transformation models are radically different, with a maximum relative error higher than 60 % at some observation points inside the low-carbon steel part.

It is also possible to numerically simulate all the transduction processes involved in the ultrasonic generation by an EMAT into a ferromagnetic part, that is to say, to simulate the Lorentz force, the magnetisation force and the magnetostriction effects. For example, figure 10 shows the fields at 50-mm-depth of transversal ultrasonic wave generated by the EMAT above-described with a static magnetic induction $\mathbf{B_S}$ along the x-direction with the body forces taken separately and all together:

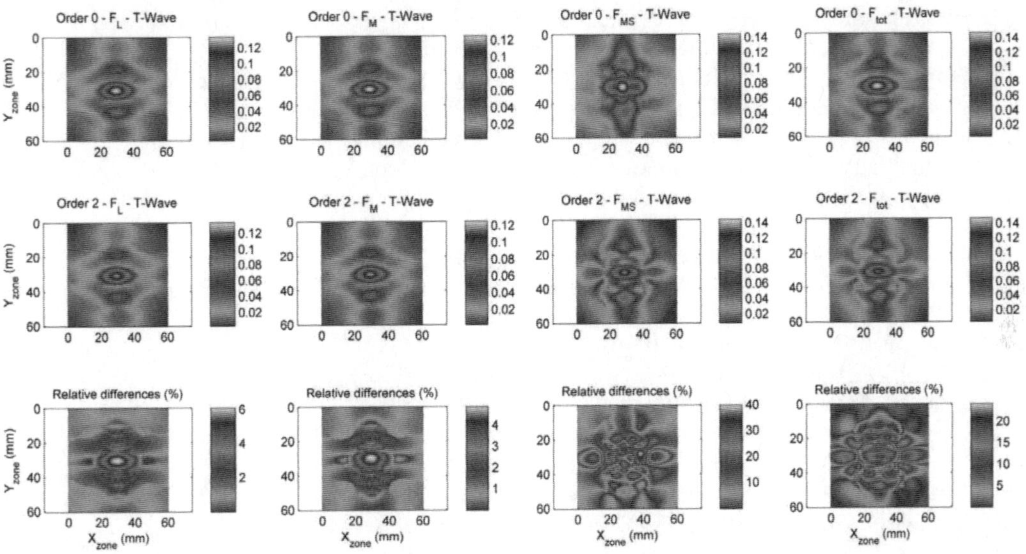

Figure 10. Maps of the maximum ultrasonic amplitude in a plane *xy* induced by all the EMAT source excitations. 1st row: 0-th order, 2nd row: 2nd order, 3rd row: relative difference. 1st column: Lorentz force, 2nd column: magnetisation force, 3rd column: magnetostriction effect, 4th column: total field.

As we can see, in this particular configuration, all the transduction processes induced by the EMAT play a role in the ultrasonic wave it generates, included the magnetostriction effects, as the amplitudes of the fields they radiate are of the same order of magnitude. In this particular configuration, it may also be observed that both the Lorentz and the magnetisation contributions look very much the same over the computation zone. Actually, they show very similar variations of the modulus of the amplitude in this zone, but their signs are opposite so that they interfere destructively. This result is well-known for a tangential static field (for example in [2]). As a consequence, the remaining field is essentially related to magnetostrictive effects; this is easily seen by comparing the maps corresponding to the field radiated by magnetostriction with those corresponding to the sum of all contributions.

5. Discussions and conclusions

The work presented herein provides a rigorous mathematical definition of an equivalent body force model for the magnetostrictive effects under a strong static magnetic field and a weak magnetoelastic coupling. This model relies on a tensorial formulation making it suitable to deal with arbitrary EMAT configuration and arbitrary geometry of the part. From the magnetostrictive equivalent body force, the derivation of the magnetostrictive equivalent surface stresses is obtained with accuracy by means of a second order expansion model, which has been numerically validated in a previous work [17]. Finally, these two mathematical models allow predicting the ultrasonic field radiated by magnetostrictive effects in ferromagnetic parts for general inspection cases in using well-established semi-analytical radiation models. Thus, all the source phenomena involved in three-dimensional configurations of EMAT and the ultrasonic field they radiate can be both efficiently and accurately predicted.

The prospects of this work consist in obtaining experimental validations of the magnetostriction model discussed here; in extending the model in the time domain in case of a strong dynamic magnetic field; and in investigating further the weak magnetoelastic coupling assumption.

References

[1] Thompson R B 1990 *Physical Acoustics* **19** 157–200
[2] Hirao M and Ogi H 2003 *EMATs for Science and Industry: Noncontacting Ultrasonic Measurements* (Boston: Kluwer Academic Publishers)
[3] Walaszek H and Bouteille P 2014 ECNDT proceedings published on-line, www.ndt.net/events/ECNDT2014/app/content/Paper/505_Walaszek.pdf
[4] Brown W F 1966 *Magnetoelastic Interactions* (Heidelberg: Springer-Verlag Berlin)
[5] Moon F C 1984 *Magneto-Solid Mechanics* (New York: John Wiley & Sons Inc)
[6] Bozorth R M 1964 *Ferromagnetism* (Hoboken: Wiley-Interscience)
[7] Lee E W 1955 *Rep. Prog. Phys.* **18** 184–229
[8] Chikazumi S 1964 *Physics of Magnetism* (New York: Wiley)
[9] Ebrahimi H, Gao Y, Kameari A, Dozono H and Muramatsu K 2013 *IEEE Trans. Magn.* **49** 1621–24
[10] Mason W 1954 *Phys. Rev.* **96** 302–10
[11] Ribichini R, Nagy P B and Ogi H 2012 *NDT &E* **51** 8–15
[12] Lamb H 1904 *Philos. Trans. R. Soc. Lond. Ser.* **A 203** 1–42
[13] Lhémery A 1994 *J. Acoust. Soc. Am.* **96** 3776–86
[14] Thompson R B 1978 *IEEE Trans. Sonics Ultrason.* **SU-25** 7–15
[15] Thompson R B 1980 *J. Nondestruct. Eval.* **1** 79–85
[16] Rouge C, Lhémery A and Ségur D 2013 *J. Acoust. Soc. Am.* **134** 2639–46
[17] Clausse B and Lhémery A 2016 *Wave Motion* **60** 135–48
[18] For detailed description of the platform, see www.extende.com

A model to predict modal radiation by finite-sized sources in semi-infinite isotropic plates

M Stévenin[1], A Lhémery[1] and S Grondel[2]

[1] CEA LIST, point courrier 120, bâtiment 565, 91191 Gif-Sur-Yvette Cedex, France
[2] IEMN-DOAE, UVHC, Le Mont Houy, 59313 Valenciennes Cedex 9, France

E-mail: alain.lhemery@cea.fr

Abstract. Elastic guided wave (GW) propagation is involved in various non-destructive testing (NDT) techniques of plate-like structures. The present paper aims at describing an efficient model to predict the GW field radiated by various sources attached at a distance of the straight boundary of an isotropic plate, a configuration often encountered in typical examinations. Since the interpretation of GW propagation and scattering in plates is made easier by the use of modal description, the model is derived in the classical theoretical framework of modal solutions. Direct radiation by a uniform source of finite size in an isotropic plate can be efficiently modelled by deriving Fraunhofer-like approximation. A rigorous treatment is proposed based upon i) the stationary phase method to describe the field after reflection at a plate edge, ii) on the computation of modal reflection coefficients for an arbitrary incidence relative to the edge and iii) on the Fraunhofer approximation to account for the finite size of the source. The stationary phase method allows us to easily express the amplitude of reflected modes, that is to say, the way waves spread, including reflections involving mode conversions. The computation of modal reflection coefficients for plane GW at oblique incidence was recently treated in the literature and our work for this very problem simply consisted in adapting it to the SAFE calculation we use to compute modal solutions.
The overall computation of the direct and reflected contributions is numerically very efficient. Once the total field is computed at a given frequency, the time-dependent field is obtained by simple Fourier synthesis.

1. Introduction

Nondestructive testing (NDT) of plate-like structures using elastic guided waves (GW) is the context of our work. Thanks to the ability of GW to propagate at long distances, one can perform fast inspections of large structures without the need for moving a transmitter and a receiver. When GW are generated by transducers of finite size, diffraction effects occur. Plate-like structures under examination are also of finite size so that GW generated in the plate can be reflected onto its edges. Since GW propagate as a series of differing modes, reflection of one GW at an edge can lead to mode-conversion phenomena.

Simulation is most helpful to help managing the complexity of GW propagation and scattering. To be realistic, transducer diffraction effects must be accounted for to accurately predict the wave-field radiated into the plate. The present work aims at proposing a model for predicting the field radiated by a finite-sized transducer into a plate of finite size, by considering as an elementary solution both the direct field radiated and the field after reflection at one straight edge of the plate. In the literature, some models [1] which can be compared to Fraunhofer-like approximations have been developed for elastic guided waves in the case of isotropic and homogeneous plates, leading to fast computation of typical diffraction effects. In the Section 3.1, we present briefly the principle of Fraunhofer-like approximation. Thanks to the stationary phase method, an equivalent expression is obtained for the reflected field, which accounts for possible mode-conversion in the reflection process. The reflection at an edge of a plane GW are the subject of a few publications [2, 3] that we will use to calculate the reflection coefficients.

Finally simulation results are shown to illustrate the capabilities of our model.

2. General expression of the displacement field

We consider a semi-infinite plate as shown in Fig. 1, of thickness $2d$, which upper plane is at $z = d$ and lower plane at $z = -d$. The plate is assumed to be made of an isotropic material. The plate edge is straight and belongs to the plane defined by the equation $x = 0$.

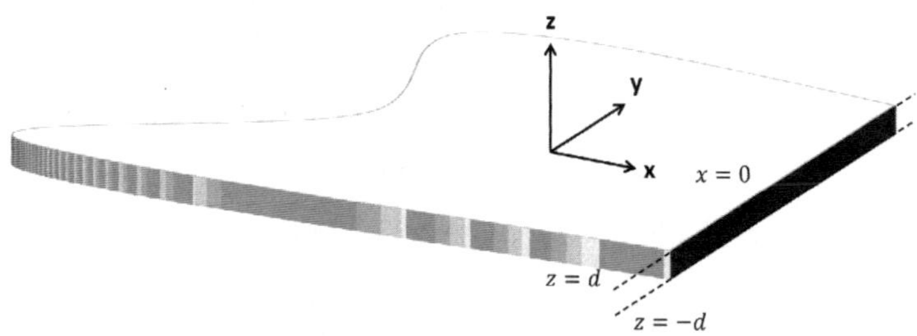

Figure 1. System geometry.

The total field $\mathbf{u}$ radiated by the source is expressed in what follows as the sum of the direct field $\mathbf{u}_d$ and the reflected field $\mathbf{u}_r$.

2.1. Direct field

The direct field radiated by a finite-sized source can be expressed in the form of a spatial convolution integral of a Green's function $\mathbf{g}_d(x, y, z)$ with a source term $\mathbf{q}(x, y)$:

$$\mathbf{u}_d(x, y, z, \mathbf{q}) = \iint_S \mathbf{g}_d(x - x', y - y', z)\mathbf{q}(x', y')dx'dy'. \tag{1}$$

The expression of this Green's function can be found in the literature [4]. By introducing its expression into equation (1) the direct field in an isotropic plate can be expressed as a sum over the propagative modes,

$$\begin{aligned}\mathbf{u}_d(x, y, z, \mathbf{q}) = \sum_m \iint_S \Bigg[& \frac{|k_m|}{2\pi}\frac{i\omega}{4P_m}\mathbf{V}_m(\varphi', z)\mathbf{V}_m(\varphi', d)^{*T}\mathbf{q}(x', y')\sqrt{\frac{2\pi}{|k_m|}} \\ & \times e^{i\frac{\pi}{4}sgn(-k_m)}e^{ik_m\sqrt{(x-x')^2+(y-y')^2}}\left(\sqrt{(x-x')^2+(y-y')^2}\right)^{-1/2}\Bigg] dx'dy'.\end{aligned} \tag{2}$$

In this expression, k_m is the wave number of the mode m,
φ' is such that $\tan\varphi' = \frac{y-y'}{x-x'}$,

P_m is the average flow power of the mode m,
and

$$\mathbf{V}_m(\varphi', z) = \mathbf{A}^{-1}(\varphi')\mathbf{W}^{(m)}(z),$$

is the displacement field distribution $\mathbf{W}^{(m)}$ for the mth mode in the direction of propagation φ', where $\mathbf{A}$ is a rotation matrix allowing us to express the field in the plate coordinates,

$$\mathbf{A} = \begin{pmatrix} \cos\varphi' & \sin\varphi' & 0 \\ -\sin\varphi' & \cos\varphi' & 0 \\ 0 & 0 & 1 \end{pmatrix}.$$

The modal contributions are computed by means of the Semi Analytic Finite Element (SAFE) method [5].

2.2. *Reflected field*

In this section, the Green's function g_r for the reflected field is considered.

$$\mathbf{u}_r(x, y, z, \mathbf{q}) = \iint_S g_r(x - x', y - y', z)\mathbf{q}(x', y')dx'dy'. \tag{3}$$

As already done when considering the direct field, the reflected field $\mathbf{u}_r$ is written as a sum over the different modal paths,

$$\mathbf{u}_r(x, y, z, \mathbf{q}) = \sum_m \sum_n \iint_S g_{rm,n}(x - x', y - y', z)\mathbf{q}(x', y')dx'dy', \tag{4}$$

where the subscript m denotes the mode considered for incident paths and n denotes the mode considered for reflected paths.

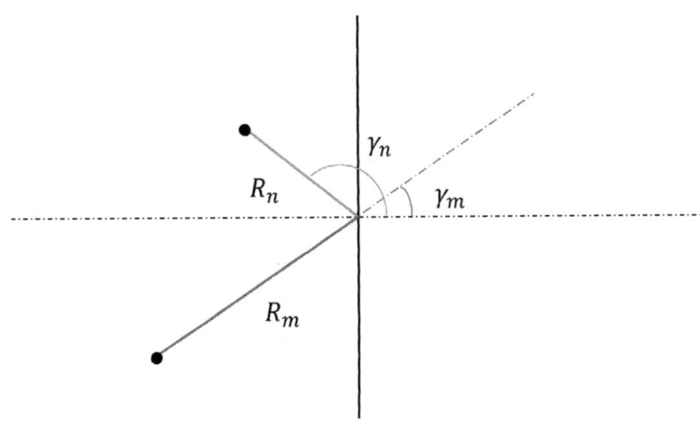

Figure 2. Notations.

Now, as point sources radiate a wavefield where waves spread with a cylindrical symmetry, wave amplitude decreases asymptotically as $r^{-1/2}$ while their phase linearly depends on the distance from the source point. At this stage, to deal with the reflection upon the straight edge, it is interesting to decompose such a wavefield as a continuous spectrum of plane waves expressed in relation to the edge position. To do this, we use Equation (28a) from Stamnes and Eide [6] and Equation (5.3a) from Stamnes [7], and after some algebra, we can obtain that

$$\sqrt{\frac{1}{kR}}e^{ikR} = \sqrt{\frac{i}{2\pi}}\int_{-\infty}^{\infty}\frac{e^{ikR}}{k_x}dk_y = \frac{1}{2\pi}\int_{-\infty}^{\infty}\frac{\sqrt{2i\pi}}{k_x}e^{ikR}dk_y. \tag{5}$$

So, a spectral elementary component of the wavefield from a point source has the following value $\frac{\sqrt{2i\pi}}{k_x}$. The Green's function expressing the reflected field along one of the paths m, n may be written as,

$$g_{rm,n} = \frac{1}{2\pi}\int G_{rm,n}\frac{\sqrt{2i\pi}}{k_{mx}}e^{i(k_{mx}(x_{m,n}-x')+k_{nx}(x-x_{m,n})+k_y y)}dk_y, \tag{6}$$

where $G_{rm,n}$ is an amplitude term which will be determined later.

$$g_{rm,n} = \sqrt{\frac{i}{2\pi}}\int G_{rm,n}\frac{e^{i(k_{mx}(x_{m,n}-x')+k_{nx}(x-x_{m,n})+k_y y)}}{k_{mx}}dk_y. \tag{7}$$

The Green's function $g_{rm,n}$ is now calculated thanks to the stationary phase method by expanding at second order the phase function about the direction of stationary-phase. For this, the second order derivative of the phase function must be calculated. The phase term has the following form,

$$f(k_y) = k_{mx}\left(x_{m,n} - x'\right) + k_{nx}\left(x - x_{m,n}\right) + k_y y. \tag{8}$$

The second order Taylor's expansion about the path of stationary phase is

$$f(k_y) = f(k_{ys}) + \left.\frac{\partial f}{\partial k_y}\right|_{k_{ys}}(k_y - k_{ys}) + \frac{1}{2}\left.\frac{\partial^2 f}{\partial k_y^2}\right|_{k_{ys}}(k_y - k_{ys})^2, \tag{9}$$

In this expression, the first order derivative vanishes as the expansion is related to the path of the stationary phase. The second order derivative is the following,

$$\begin{aligned}\left.\frac{\partial^2 f}{\partial k_y^2}\right|_{k_{ys}} &= -\frac{(x_{m,n}-x')}{k_{mx}}\frac{1}{\cos^2\gamma_m} - \frac{(x-x_{m,n})}{k_{nx}}\frac{1}{\cos^2\gamma_n} \\ &= -\frac{1}{k_m}\frac{1}{\cos^2\gamma_m}\left(R_m + R_n\frac{k_m}{k_n}\frac{\cos^2\gamma_m}{\cos^2\gamma_n}\right),\end{aligned} \tag{10}$$

where, if $(x_{m,n}, y_{m,n})$ denotes the point of the plate edge where reflection arise, R_m is the distance of this point to the source point.
Likewise, R_n is the distance between the calculation point and the point where reflection arises.

γ_m and γ_n are the phase directions before and after reflection, respectively. We can now replace the phase with its development,

$$\begin{aligned}&\int_{-\infty}^{\infty}\frac{e^{i(k_{mx}(x_{m,n}-x')+k_{nx}(x-x_{m,n})+k_y y)}}{k_{mx}}dk_y \\ &= \frac{e^{i(k_m R_m + k_n R_n)}}{k_{mx}}\int_{-\infty}^{\infty}e^{-i\frac{1}{2}\left(\frac{1}{k_m}\frac{1}{\cos^2\gamma_m}\left(R_m+R_n\frac{k_m}{k_n}\frac{\cos^2\gamma_m}{\cos^2\gamma_n}\right)\right)(k_y-k_{ys})^2}dk_y\end{aligned} \tag{11}$$

We recognize in this expression a Gaussian integral,

$$\begin{aligned}&\int_{-\infty}^{\infty}e^{-i\frac{1}{2}\left(\frac{1}{k_m}\frac{1}{\cos^2\gamma_m}\left(R_m+R_n\frac{k_m}{k_n}\frac{\cos^2\gamma_m}{\cos^2\gamma_n}\right)\right)(k_y-k_{ys})^2}dk_y \\ &= e^{-i\frac{\pi}{4}\operatorname{sgn}\left(\frac{1}{k_m}\left(R_m+R_n\frac{k_m}{k_n}\frac{\cos^2\gamma_m}{\cos^2\gamma_n}\right)\right)}\sqrt{\frac{2\pi|k_m|^2\cos^2\gamma_m}{\left|R_m k_m + R_n k_n\left(\frac{k_m}{k_n}\frac{\cos\gamma_m}{\cos\gamma_n}\right)^2\right|}}\end{aligned} \tag{12}$$

So we have,

$$\int_{-\infty}^{\infty} \frac{e^{i(k_{mx}(x_{m,n}-x')+k_{nx}(x-x_{m,n})+k_y y)}}{k_{mx}} dk_y = e^{i(k_m R_m + k_n R_n)} e^{-i\frac{\pi}{4}\operatorname{sgn}\left(\frac{1}{k_m}\left(R_m + R_n \frac{k_m}{k_n}\frac{\cos^2\gamma_m}{\cos^2\gamma_n}\right)\right)} \sqrt{\frac{2\pi}{\left|R_m k_m + R_n k_n \left(\frac{k_m}{k_n}\frac{\cos\gamma_m}{\cos\gamma_n}\right)^2\right|}} \tag{13}$$

$$g_{rm,n} = \sqrt{\frac{i}{2\pi}} e^{i(k_m R_m + k_n R_n)} e^{-i\frac{\pi}{4}\operatorname{sgn}\left(\frac{1}{k_m}\left(R_m + R_n \frac{k_m}{k_n}\frac{\cos^2\gamma_m}{\cos^2\gamma_n}\right)\right)} \times G_{rm,n} \sqrt{\frac{2\pi}{\left|R_m k_m + R_n k_n \left(\frac{k_m}{k_n}\frac{\cos\gamma_m}{\cos\gamma_n}\right)^2\right|}}. \tag{14}$$

Just before the reflection, the displacement field due to the m mode is given by,

$$\mathbf{u}_{dm}(x_p, y_p) = \left[\frac{|k_m|}{2\pi}\frac{i\omega}{4P_m}\mathbf{V}_m(\varphi_m, z)\mathbf{V}_m(\varphi_m, d)^{*T}\mathbf{q}(x', y') \times \sqrt{\frac{2\pi}{|k_m R_m|}} e^{i\frac{\pi}{4}sgn(-k_m)} e^{ik_m R_m}\right], \tag{15}$$

so that just after the reflection, the displacement field can be written as

$$\mathbf{u}_{rm,n}(x_p, y_p) = r_{m,n}\frac{|k_m|}{2\pi}\frac{i\omega}{4P_m}\mathbf{V}_n(\varphi_n, z)\mathbf{V}_m(\varphi_m, d)^{*T}\mathbf{q}(x', y') \times \sqrt{\frac{2\pi}{|k_m R_m|}} e^{i\frac{\pi}{4}sgn(-k_m)} e^{ik_m R_m}. \tag{16}$$

where, φ_m is the angle between the point where reflection arises and the source point relative to the edge normal. Likewise, φ_n is the angle between the calculation point and the reflection point.

Our goal is now to calculate $G_{m,n}$. The reflected wave n has the following form,

$$\mathbf{u}_r(x, y) = \sum_m \sum_n a_{m,n}\mathbf{V}_n(z). \tag{17}$$

Just before the reflection, for one path, the displacement field due to the incident mode m is given by,

$$\mathbf{u}_{dm}(x_p, y_p) = \left[\frac{|k_m|}{2\pi}\frac{i\omega}{4P_m}\mathbf{V}_m(\varphi_m, z)\mathbf{V}_m(\varphi_m, d)^{*T}\mathbf{q}(x', y') \times \sqrt{\frac{2\pi}{|k_m R_m|}} e^{i\frac{\pi}{4}sgn(-k_m)} e^{ik_m R_m}\right], \tag{18}$$

so just after the reflection, the displacement field can be expressed as

$$\mathbf{u}_{rm,n}(x_p, y_p) = \left[r_{m,n}\frac{|k_m|}{2\pi}\frac{i\omega}{4P_m}\mathbf{V}_n(\varphi_n, z)\mathbf{V}_m(\varphi_m, d)^{*T}\mathbf{q}(x', y') \times \sqrt{\frac{2\pi}{|k_m R_m|}} e^{i\frac{\pi}{4}sgn(-k_m)} e^{ik_m R_m}\right], \tag{19}$$

where $r_{m,n}$ denotes the reflection coefficient related to the path m, n which will be calculated in section 4. At this stage, we have

$$G_{rm,n} = \sqrt{\frac{2\pi}{i}} r_{m,n} \frac{|k_m|}{2\pi} \frac{i\omega}{4P_m} \mathbf{V}_n(\varphi_n, z)\mathbf{V}_m(\varphi_m, d)^{*T}. \tag{20}$$

Finally, the displacement field after reflection can be expressed by:

$$\mathbf{u}_r = \sum_m \sum_n \left[r_{m,n} \frac{|k_m|}{2\pi} \frac{i\omega}{4P_m} \mathbf{V}_n(\varphi_n, z)\mathbf{V}_m(\varphi_m, d)^{*T} \mathbf{q}(x', y') \right. \\ \left. \times e^{if_{m,n}(\gamma_m)} e^{i\frac{\pi}{4} sgn\left(\frac{\partial^2 f_{m,n}}{\partial \gamma^2}|_{\gamma_m}\right)} \sqrt{\frac{2\pi}{\left|R_m k_m + \left(\frac{k_m}{k_n}\frac{\cos(\gamma)}{\cos(\gamma_n)}\right)^2 R_n k_n\right|}} \right]. \tag{21}$$

In the next section, we will explain how efficiently taking into account efficiently the finite size of the source and how determining paths between it and the calculation points.

3. Finite-sized sources

3.1. Fraunhofer-like approximation

In the isotropic case and for the direct field, the integral over the source surface can be accurately and efficiently evaluated thanks to Fraunhofer approximation. This approximation is based on a first order approximation of the phase term and a zero-th order approximation of the amplitude. Let (x_c, y_c),denote the coordinates of the source centre. The direct displacement field given by Eq. 2 under Fraunhofer approximation writes

$$\mathbf{u}_d(x, y, z, \mathbf{q}) = \left(\sqrt{(x-x_c)^2 + (y-y_c)^2}\right)^{-1/2} \sum_m \frac{i\omega}{4P_m} \sqrt{\frac{|k_m|}{2\pi}} \mathbf{V}_m(\varphi_c, z)\mathbf{V}_m(\varphi_c, d)^{*T} \\ \times e^{i\frac{\pi}{4} sgn(-k_m)} e^{ik_m\sqrt{(x-x_c)^2+(y-y_c)^2}} \iint_S \left[e^{-ik_m \frac{xx'+yy'}{\sqrt{(x-x_c)^2+(y-y_c)^2}}} \mathbf{q}(x', y') \right] dx'dy'. \tag{22}$$

The surface integral is easily calculated for standard geometries of source (rectangular or circular) generating in-plane or out-of-plane uniform stresses.

For example, considering rectangular and disk transducers generating uniform normal stress over their surface, the direct fields radiated into the plate by these sources are readily given by

$$\mathbf{u}_d(x, y, z, \mathbf{q}) = \left(\sqrt{(x-x_c)^2 + (y-y_c)^2}\right)^{-1/2} \sum_m \frac{i\omega}{4P_m} \sqrt{\frac{|k_m|}{2\pi}} \mathbf{V}_m(\varphi_c, z)\mathbf{V}_m(\varphi_c, d)^{*T} \\ \times e^{i\frac{\pi}{4} sgn(-k_m)} e^{ik_m\sqrt{(x-x_c)^2+(y-y_c)^2}} F_m(x - x_c, y - y_c)\mathbf{q}(x_c, y_c). \tag{23}$$

F_m is a modal function which stands for the result of the surface integral. For a rectangular-shaped transducer of aperture $L_x \times L_y$ it is given by,

$$F_{m_{rect}}(\varphi_c) = L_x L_y \,\mathrm{sinc}\left(k_m \frac{L_x}{2} \cos\varphi_c\right) \mathrm{sinc}\left(k_m \frac{L_y}{2} \sin\varphi_c\right)$$

and for a disk-shaped transducer of radius a, by,

$$F_{m_{disk}} = \pi a^2 \frac{2J_1(k_m a)}{k_m a}.$$

As we can see, these expressions are only dependent on the angle φ_c between the source centre and the calculation point (at least in the case of a rectangular source). Since such an approximation is already valid before the reflection, it can be all the more used in the case of reflected paths taking into account the direction between the point where reflection arises and the centre of the source.

3.2. Paths between the source and the calculation points

The centre of the source is, as in the previous section (x_c, y_c). The coordinates of the calculation point are denoted by (x, y). We will search the coordinate of the reflection point $(0, y_p)$ for a couple m, n of modes.

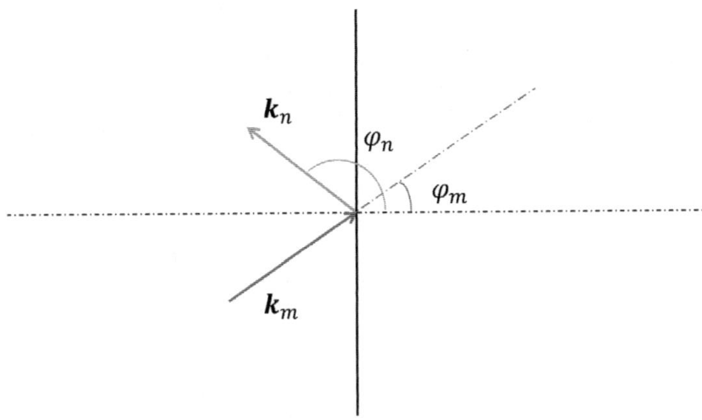

Figure 3. Reflection at an edge.

In an isotropic plate, the phase direction is the same as the energy direction. Denoting the incident angle by φ_m and the angle of reflection by φ_n, as shown on the figure 3, the Snell-Decartes' law gives us,

$$k_m \sin(\varphi_m) = k_n \sin(\varphi_n), \tag{24}$$

$$k_m \frac{y_p - y_c}{\sqrt{(y_p - y_c)^2 + x_c^2}} = k_n \frac{y - y_p}{\sqrt{(y - y_p)^2 + x^2}}, \tag{25}$$

After some algebra, we obtain the following polynomial to solve:

$$a_4 y_p^4 + a_3 y_p^3 + a_2 y_p^2 + a_1 y_p + a_0 = 0, \tag{26}$$

with

$$\begin{aligned}
a_4 &= \left(k_m^2 - k_n^2\right) \\
a_3 &= -2(k_m^2 - k_n^2)(y_c + y) \\
a_2 &= (y_c^2 + 4y_c y + y^2)(k_m^2 - k_n^2) + x^2 k_m^2 - x_c^2 k_n^2 \\
a_1 &= -2\left((y_c y^2 + y_c^2 y)(k_m^2 - k_n^2) + x^2 y_c k_m^2 - x_c^2 y k_n^2\right) \\
a_0 &= (y_c y)^2 (k_m^2 - k_n^2) + (x y_c k_m)^2 - (x_c y k_n)^2 .
\end{aligned} \tag{27}$$

This polynomial has four roots. If a reflection point exists, it is the real solution where the value of y_p is between those of y and y_c. Now if we have a reflection point for a couple m, n of modes, we know the path for these between the source centre and the calculation point. For each of these paths, we must now calculate the reflection coefficients.

4. Reflection coefficients

To calculate the reflection coefficients, we use the method proposed by Santhanam and Demirli [2]. In this method, stress free boundary conditions are assumed at the plate edge. For an incident mode m and all the reflected modes n, we can write the following system:

$$\begin{cases} \sigma_{11}^{(m)} + \sum_{n=1}^{\infty} r_{m,n}\sigma_{11}^{(n)} = 0 \\ \\ \sigma_{12}^{(m)} + \sum_{n=1}^{\infty} r_{m,n}\sigma_{12}^{(n)} = 0 \\ \\ \sigma_{13}^{(m)} + \sum_{n=1}^{\infty} r_{m,n}\sigma_{13}^{(n)} = 0 \end{cases} . \tag{28}$$

In this system, all the stress tensors $\sigma^{(i)}$ are written in the plate coordinate system. The SAFE method allows us to compute them in the propagation system. Therefore, in order to express them in the plate coordinate system, a transformation matrix must be introduced, as presented by Gunawan and Hirose [3]. The change of coordinate system is written as follows,

$$\sigma = Q\sigma_p Q^t, \tag{29}$$

where Q denotes the transformation matrix, σ_p denotes the stress expressed in the coordinate system attached to the wave propagation considered, and σ denotes the stress expressed in the plate coordinate system. The transformation matrix is readily given by:

$$Q = \begin{pmatrix} C & S & 0 \\ -S & C & 0 \\ 0 & 0 & 1 \end{pmatrix} . \tag{30}$$

We can see on the figure 3 that the incident mode makes an angle φ_m with respect to the plate frame. For this mode,

$$\begin{aligned} C_m &= \cos\varphi_m \\ S_m &= -\sin\varphi_m \end{aligned} \tag{31}$$

Now, the transformation matrices for the different reflected modes are similarly obtained by introducing the following notations:

$$\xi = k_m \sin\varphi_m \tag{32}$$

$$\mu = \pm\sqrt{k_n^2 - \xi^2} \tag{33}$$

$$C_n = -\frac{\mu}{k_n} \tag{34}$$

$$S_n = -\frac{\xi}{k_n} \tag{35}$$

Choose the sign in the Eq. (33) is made according to [3]. It must be noticed that for propagative modes, C_n an S_n are real valued. In these cases, Q is nothing but a rotation matrix and solving Eq. (26) leads to the obtaining of a reflection point.

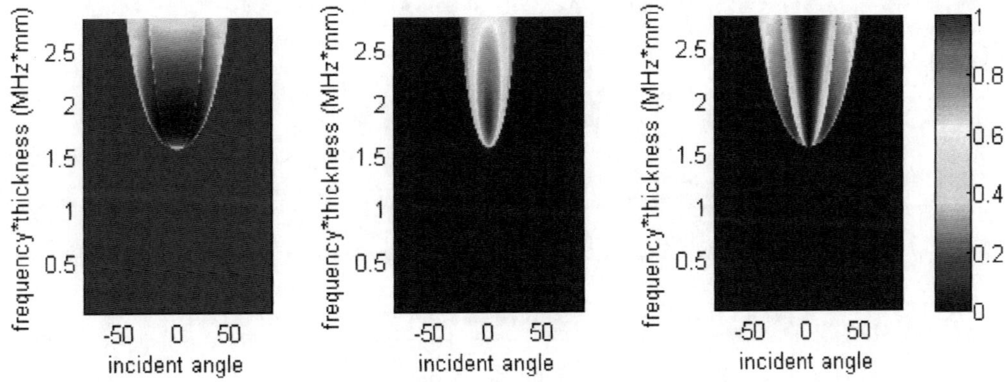

Figure 4. Energy reflection coefficient for A0 incident and for (from left to right) A0, A1, SH1 reflected.

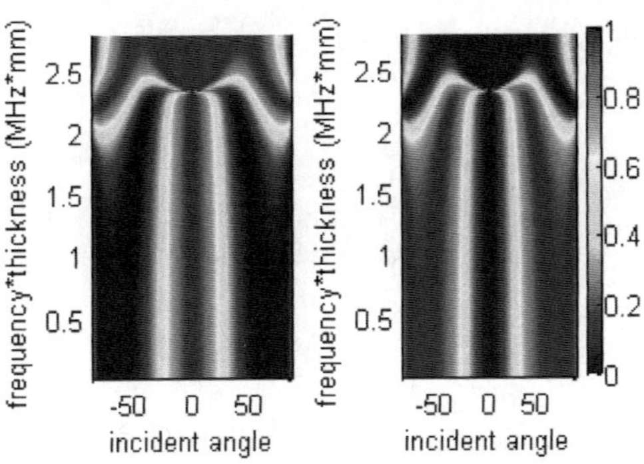

Figure 5. Energy reflection coefficient for S0 incident and for (from left to right) S0, SH0 reflected.

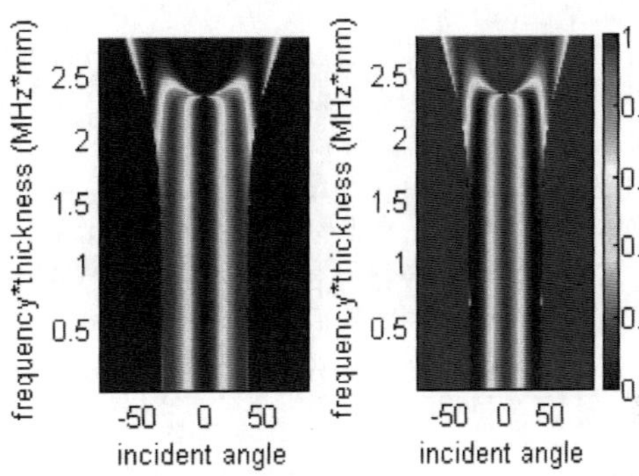

Figure 6. Energy reflection coefficient for SH0 incident and for (from left to right) S0, SH0 reflected.

Once all the stress tensors are written in the same coordinate system, we can calculate the $r_{m,n}$ coefficients appearing in Eq. (28), by truncating the infinite sum over a finite number N of modes. All the propagative modes must be accounted for, together with a finite number of inhomogeneous and evanescent modes.

Energy reflection coefficients $\mathcal{R}_{m,n}$ can also be computed likewise by means of,

$$\mathcal{R}_{m,n} = \frac{P_n}{P_m}(r_{m,n} r^*_{m,n}), \tag{36}$$

where P denotes the mean energy flux of the considered mode along the plate.

Figures 4, 5 and 6 present the energy reflection coefficients for the three first propagative modes as functions of the incident angle and of the frequency-thickness product. The plate considered in this simulation is made of aluminium and is 3 mm thick. Dispersion curves of propagative modes are plotted on Fig. 7.

Below the first cut-off frequency, the A0 mode is the only one propagative mode to present an antisymmetric shape. That is why an incident A0 mode is totally reflected as another A0 mode.

IOP Conf. Series: Journal of Physics: Conf. Series **797** (2017) 012005 doi:10.1088/1742-6596/797/1/012005

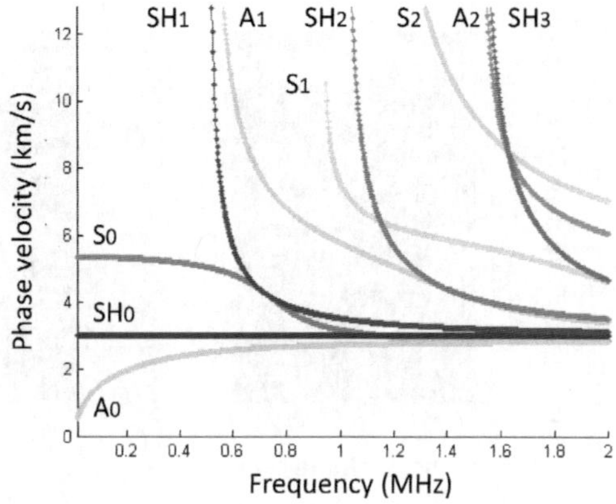

Figure 7. Dispersion curves of a 3-mm-thick aluminium plate.

Both the S0 and the SH0 modes have a symmetric shape. Therefore, when a S0 or a SH0 mode interacts with the free edge of the plate, mode conversion occurs.

5. Results in time domain

Throughout the theoretical derivations of field models described in previous sections, a single frequency excitation was assumed. In what follows, time-dependent wavefields are presented which are obtained by standard Fourier synthesis.

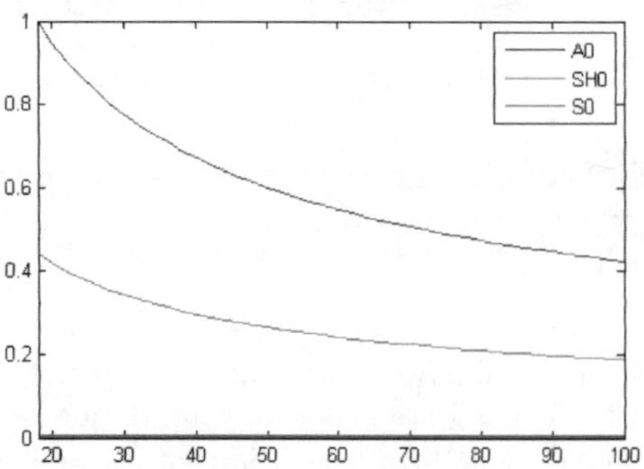

Figure 8. Normalized normal modal displacement along an axis from a distance of λ_{max} of the edge of the source centered at $(0, 0)$.

At first, we will compare the normal displacement due to each mode at the frequency of 400 kHz. As presented on Fig. 8, the A0 mode and the S0 mode are predominant while the SH0 mode is negligible.

In this simulation, we consider a 5-mm-diameter circular transducer centered at $(-100, 0)$ which generates a normal uniform stresses excited by a toneburst of sine wave of 400 kHz centre frequency weighted by a Hanning window. We present on figure 9 the propagation of the waves with a reflection at four different times.

The first instant $t = 18\mu s$ (a) was chosen before any reflection phenomenon occurs; calculation points reached by some acoustic energy are insonified only by direct contributions from the source. The second instant $t = 31\mu s$ (b) was chosen at the beginning of the reflection process, when only the fastest mode reached the plate edge; interferences of direct and reflected contributions are visible close to the edge. The third instant $t = 54\mu s$ (c) was chosen just after the slowest

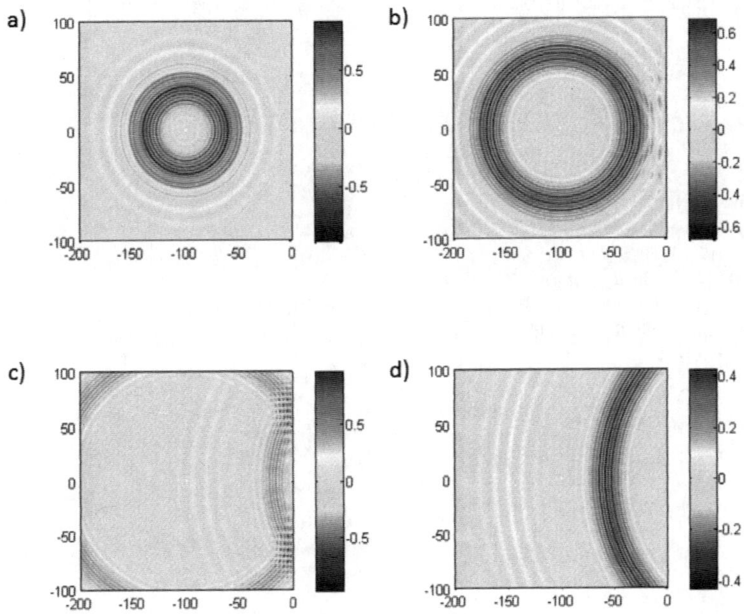

Figure 9. Four snapshots of the normal displacement associated to guided wave propagation in a plate with account of edge reflection (a) at $18\mu s$, (b) at $31\mu s$, (c) at $54\mu s$, and (d) at $72\mu s$.

contributions finished to reflect on the edge; at some calculation points, interferences with direct contributions are still visible. The last instant $t = 72\mu s$ (d) was chosen when all the points of the computation zone are only insonified by waves reflected on the plate edge; there is no more interference.

The S0 mode is the fastest mode. So, it is the first to be reflected. Moreover, as expected we distinguish only the contribution of two modes.

6. Conclusion

An efficient model has been proposed to compute the elastic guided wave field radiated by finite-sized sources in a semi-infinite isotropic plate. The model gives the field at a point of the plate as the superposition of direct contributions from the source with contributions involving reflections onto the free edge of the semi-infinite plate. Modes are computed by means of the SAFE method but classical formulas for Lamb waves and transverse horizontal guided waves could be easily introduced. The calculation of reflected contributions is based upon the stationary phase method, stationary phase paths being easily determined by solving a simple polynomial equation. Fraunhofer-like approximation is used for efficiently evaluate the surface integral over transducer surface. This approximation was already known to combine efficiency and accuracy for the field radiated into an infinite plate [8]; this is all the more true for longer wavepaths involving edge reflection.

The recent article of Feng *et al.* [9] develops the mode-matching method for computing the scattering of obliquely incident guided waves by a straight feature. This method can be used straightforwardly in our model to deal with more complex boundary conditions at plate edges. Likewise, the integration along energy directions developed for predicting transducer diffraction effects in anisotropic plates [8], could be adopted to generalize the present model to anisotropic cases.

References

[1] Raghavan A and Cesnik C E S 2005 *Smart Mater. Struct.* **14** 1448–1461
[2] Santhanam S and Demirli R 2013 *Ultrasonics* **53** 271–282
[3] Gunawan A and Hirose S 2007 *Mater. Trans.* **48** 1236–1243
[4] Velichko A and Wilcox P D 2007 *J. Acoust. Soc. Am.* **121** 60–69
[5] Taupin L, Lhémery A and Inquiété G 2011 *J. Phys.: Conf. Ser.* **269** 012002
[6] Stamnes J J and Eide H A 1998 *J. Opt. Soc. Am. A* **15** 1285–1291
[7] Stamnes J J 1986 *Waves in focal regions* (CRC Press)
[8] Stévenin M, Lhémery A and Grondel S 2016 *J. Phys.: Conf. Ser.* **684** 012004
[9] Feng F, Shen Z and Shen J 2016 *Wave Motion* **60** 84–94

IOP Conf. Series: Journal of Physics: Conf. Series **797** (2017) 012006 doi:10.1088/1742-6596/797/1/012006

Noise of sliding rough contact

Alain Le Bot

Ecole centrale de Lyon, CNRS LTDS UMR 5513, 36, av. Guy de Collongue 69134 Ecully, France

E-mail: alain.le-bot@ec-lyon.fr

Abstract. This article is a discussion about the origin of friction noise produced when rubbing solids having rough surfaces. We show that noise emerges from numerous impacts into the contact between antagonist asperities of surfaces. Prediction of sound sources reduces to a statistical problem of contact mechanics. On the other hand, contact is also responsible of dissipation of vibration. This leads to the paradoxical result that the noise may not be proportional to the number of sources.

1. Introduction

Friction of solids having rough surfaces produces a typical wideband noise with generally a low sound level. This is for example the case when rubbing hands or pushing a small object on a table. Other examples are the sounds produced by a piece of sandpaper or a scouring sponge on a saucepan. In all these examples, the contact is weak and mechanical interaction is confined to the interface region at the top of surface asperities. All these sounds may be referred to as roughness noise [1].

This study focuses on the question of the mechanical origin of roughness noise. We present simple experiments and numerical simulations which aim to clarify what happens into the contact and leads to the noise emission. Discussion is conducted in the case of flat metal samples whose surface have been unpolished.

2. Acoustic characterization of roughness noise

A simple way to produce a roughness noise is to rub against each other two nominally flat pieces of metal (figure 1), such as proposed in Ref. [2]. Surfaces have been prepared by electric-discharge machining which leads to a random topography. Figure 2 shows the probability density function of heights of asperities of the samples. The values of quadratic roughness (standard deviation), Skewness and Kurtosis are specified. The Kurtosis is close to 3, the theoretical value for a Gaussian distribution. The distribution is clearly Gaussian, homogeneous and isotropic.

Several remarks must be pointed out concerning noise emission.

The sound level increases with sliding velocity as it can easily be checked with the metal samples (try it rubbing your hands!). The evolution law is of exponential type [3]

$$\Delta L_{\rm p} = 20\log_{10}(V_2/V_1)^{\beta} \tag{1}$$

where $\Delta L_{\rm p}$ is the difference of sound pressure level (dB) between speeds V_1 and V_2.

Figure 1. Simple experiment of friction sound with two steel sample with surfaces prepared by electric-discharge machining.

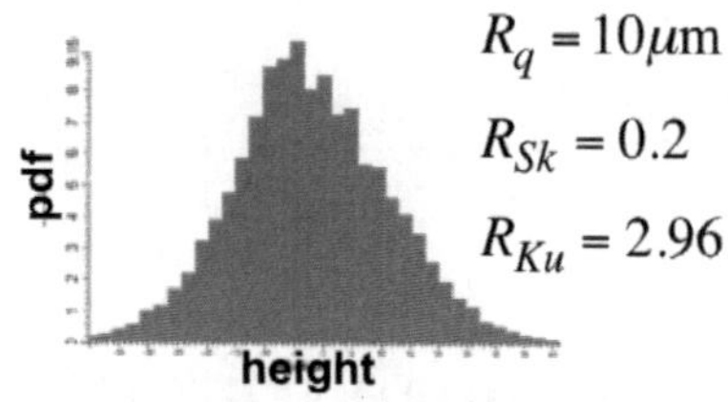

Figure 2. Distribution of asperity heights.

The sound level also increases with roughness. An accurate measurement of the dependence of sound with roughness leads again to an exponential law [4, 3]

$$\Delta L_{\rm p} = 20\log_{10}(R_2/R_1)^{\alpha} \tag{2}$$

where $\Delta L_{\rm p}$ is the difference of sound pressure level between roughness R_1 and R_2.

Roughness noise is characterized by a relatively flat power spectrum. In figure 3 is shown the power spectral density of friction sound recorded by a microphone for two different sliding velocities. The spectrum is very wide and the entire audio frequency band is covered. This wide band nature of sound is the main difference of roughness noise with other types of friction sounds such as squeal noise or other narrow band sounds (a fundamental frequency and few harmonics).

In figure 3, we can also observe some emerging frequencies. These peaks exactly match with the natural frequencies of solids without contact. The mechanical contact between the solids does not modify the natural frequencies (a strong contact stiffness would result in a shift towards high frequencies). The mechanical coupling between samples is therefore weak. This is an important characteristics of roughness noise. This condition is generally established under a light contact pressure.

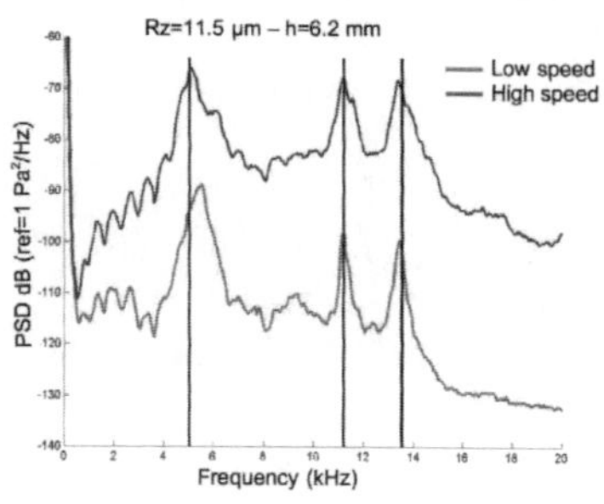

Figure 3. Power spectrum of roughness noise produced by sliding two parallelepipedic metal pieces. Red, low sliding speed; blue, high sliding speed. The vertical black lines indicates the natural frequencies of solids.

3. Mechanical origin of roughness noise

The mechanical process responsible of roughness noise must be discovered at the scale surface asperities. Surface asperities have a typical size of order of dozen of micrometres. They represent obstacles against motion. Percussion at the top of antagonist asperities act on solids as short and light impacts but at a very high rate. Since the surface is random by nature, these impacts are disordered events (Fig. 4).

Figure 4. Mechanical origin of friction noise. Percussion of asperities produces vibration responsible of noise emission.

These micro-impacts behave like small hammer strokes which mechanically excite the structure. Since the Fourier transform of a pulse is constant, we now better understand why noise spectrum is so wide. More accurately, we may assess the impact duration for instance with a model of an elastic sphere impacting a rigid plane. For a steel-steel contact, the impact duration is of order of 0.1 ms. The frequency bandwidth of sound is therefore larger than 10 kHz (we measured sound up to 50 kHz with a 1/4" condenser microphone).

The rate of impacts may also be assessed by as follows. The length of correlation in the contact may be defined as the smallest distance of motion for which the spot population is entirely re-newed (Fig. 5). This definition was given by Rabinowicz to explain the transition between static and kinetic friction [5]. Rabinowicz measured this distance D_0 by an indirect technique and obtained that it ranges from 1 to 10 μm for different metallic contacts.

In static condition, the number of contact spots is generally low say, of order of $n_0 \sim 10$ which corresponds to a bearing rate of few percents. But in sliding condition with a speed V, under the assumption of quasi-static movement, the population is entirely re-newed after a time D_0/V. The impact rate is therefore

$$n = n_0 \frac{V}{D_0} \sim 10^5 \quad \text{impacts/s} \tag{3}$$

for a typical sliding speed $V \sim 10$ cm/s and $D_0 = 10$ μm.

A so large impact rate $n \sim 10^5$ combined with a short duration $\tau \sim 10^{-4}$ s leads to a large overlap of impacts (say dozens of impacts at the same time). It is therefore impossible to ear (or measure) separately the different impacts. The fact that roughness noise is perceived as a steady-state noise by the human ear stems from microscopic transient events but so numerous that they appear as a permanent source.

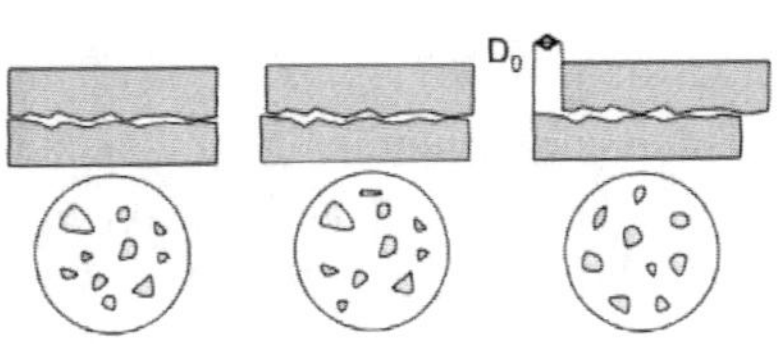

Figure 5. Correlation length of an interface. Under a sliding of length $< D_0$, the population of spots is not significantly modified. When the sliding is $> D_0$, the contact is entirely affected and new spots are created.

4. Numerical simulation of roughness noise

The numerical simulation of friction noise is possible in principle with the only laws of contact mechanics. This is a problem of elasto-dynamics in the presence of contact.

Since the impact duration is of order of 0.1 ms, the time step in numerical simulation must be much lower. In practice we used a time step of 10^{-8} s with classical time-integration schemes. Contact detection between the top of asperities requires a very fine mesh. We applied classical contact algorithms (penalty or Lagrange multipliers methods) with spatial step of 4 μm.

IOP Conf. Series: Journal of Physics: Conf. Series **797** (2017) 012006 doi:10.1088/1742-6596/797/1/012006

However, the wavelengths of sound and elastic waves in solids are usually very large (30 cm for sound at 1 kHz and several centimetres for elastic waves). So, a finite element model of the elasto-dynamics problem does not require a so fine mesh.

This contradiction between a fine mesh for contact detection and a gross mesh for vibrations highlights that roughness sound simulation is basically a multi-scale problem numerically ill-posed.

We circumvented this difficulty by projecting the equations of motion on a modal basis. If $u(x,t)$ denotes the vibration field and ω_k, $\psi_k(x)$, $k=1,2,\dots$ the natural frequencies and mode shapes, then the modal projection reads,

$$u(x,t) = \sum_{k\geq 1} U_k(t)\psi_k(x) \tag{4}$$

where U_k is the modal amplitude to determine. The equations of motion on modal amplitudes are

$$m_k\left(\ddot{U}_k(t) + 2\zeta\omega_k\dot{U}_k(t) + \omega_k^2 U_k(t)\right) = F_k(t) \tag{5}$$

where m_k is the modal mass, ζ the modal damping factor and F_k the modal force induced by contact. The number of equations of motion is very low since relatively few modes are involved in this problem. But computing modal contact forces F_k requires to detect all contact points at each time step. This is achieved by computing the surface deformation on a fine mesh at each time step. This is of course the most time consuming step. The gain compared with classical finite element method is therefore in the time-integration scheme limited to few equations of motion. Details of the method may be found in Ref. [6].

An example of result is shown in figures 6, 7. The studied system is composed of a small cube of length 2 cm made of stainless steel sliding on a large elastic plate of thickness 2 mm. The surfaces have been modelled by numerically generating a gaussian surface of various roughnesses. We can observe in figure 6 that the vibrational level (and therefore the Sound Pressure Level) follows a log evolution with sliding velocity and roughness. This is in agreement with empirical laws of Eqs. (1) and (2). Furthermore, in figure 7, we also observe that the rate of impacts is about 10^5 impacts/s for $V = 10$ cm/s in a fine agreement with Eq. (3). However, it also appears that the impact rate decreases with sliding speed. This may be interpreted as a dynamical effect. Impacts are more strong when the movement is more rapid. This may result in a loss of contact and an increase of flight duration. This phenomenon was not taken into account in Eq. (3) based on quasi-static assumption.

5. Dissipation of vibration

It is well-known in engineering that contacts are responsible of strong dissipation in vibrations. A metallic structure, such as a shell or a frame, is generally highly reverberant and may easily transmit sound and vibration in the overall system. However, the same structure attached by bolts, welding points, or rivets presents a much lower reverberation time. The presence of contact in attachments induces a strong dissipation of vibration by micro-slips into the contact. This mechanism is sometimes used to design low reverberant components in built-up structures. A particularly impressive example is car oil sump made by stamping together two metal sheets instead of a single one. This gives a highly absorbant element which does not contribute to radiate sound.

The following experiment illustrates this phenomenon [7]. We measured the sound pressure level with a sonometre, induced by the sliding of sugar lumps on a table surface (large wood table). Not surprisingly, we observed that the sound pressure level increases with the number of sugar lumps. Sugar lumps are sound sources that may be considered as uncorrelated. Their powers may be summed and therefore the theoretical law is an increase by 10 dB of Sound

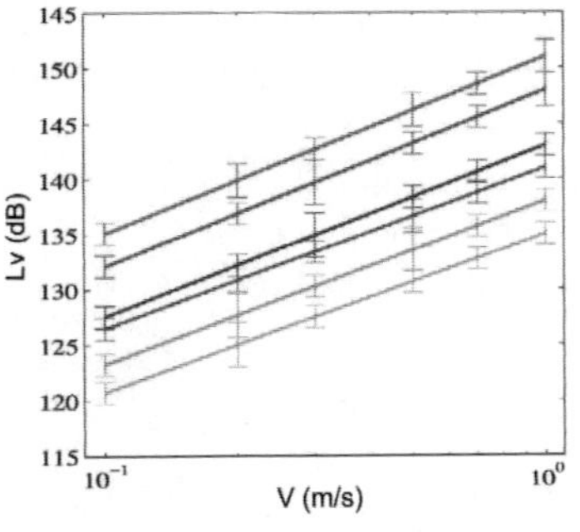

Figure 6. Vibrational level versus sliding speed for various roughnesses (from bottom to top Ra=3, 5, 8, 10, 20, 30 μm).

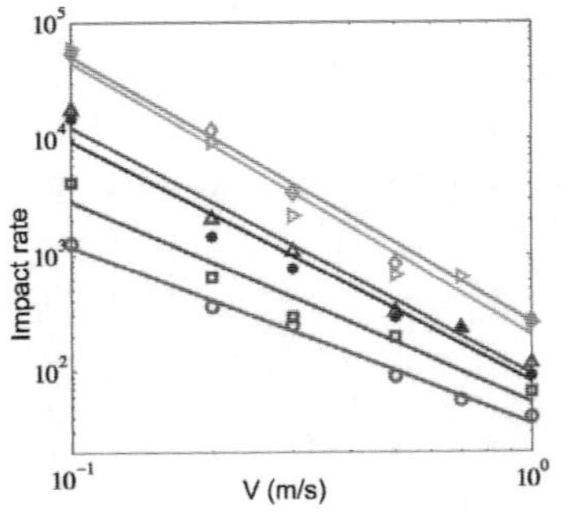

Figure 7. Impact rate versus sliding speed for various roughnesses (from top to bottom Ra=3, 5, 8, 10, 20, 30 μm).

Pressure Level by decade of sugar lumps. The overall increase of 18 dB for a number of sugar lumps ranges from 1 to 100 indicates therefore an experimental slope of 9 dB per decade which is very close to this theoretical value.

However, when reproducing this simple experiment on a drum membrane, we observed that the Sound Pressure Level is almost constant (we did the experiment from 1 to 80 sugar lumps). This result is rather surprising at first sight. If the lumps are uncorrelated noise sources on a table wood, we may wonder why they are not on a drum membrane.

The key of this phenomenon is not in the additive law of sound power but rather in dissipation process. We may propose the following thermodynamical reasoning. The structure (table plate or drum membrane) is a tank of vibrational energy. Sources (sugar lumps) inject vibrational energy in this tank. The power balance reads $P_{\text{inj}} = P_{\text{diss}}$ where P_{inj} is the power being injected by sources and P_{diss} the dissipated power. The total power supplied by sources is $P_{\text{inj}} = Np$ where N is the number of sources and p the unit power. For dissipation of vibration in materials, a widely-accepted dissipation law [8] is

$$P_{\text{diss}} = \eta\omega WA \tag{6}$$

where η is the damping loss factor, ω the central frequency, W the energy per unit area (assumed to be uniformly distributed), and A the plate area. Substituting P_{inj} and P_{diss} into the energy balance gives $Np = \eta\omega WA$. The energy density is proportional to the number of sources, $W \propto N$. This proportionality of vibrational energy and therefore the radiated sound with the number of sources explains the slope of 10 dB per decade observed on wood table.

But in the case of membrane, natural dissipation of vibrations is almost negligible. In the sugar lumps experiment, dissipation of vibrations mainly stems from the presence of lumps themselves. Micro-slips occurring at the interface between the sugar lumps and membrane are responsible of most of dissipation. It is therefore reasonable to assert that dissipation is proportional to contact area. More specifically, we may infer by analogy with Eq. (6) that

$$P_{\text{diss}} = \eta'\omega WS \tag{7}$$

where η' is an ad-hoc damping loss factor (due to contact) and S the total contact area. Letting $S = Ns$ where s is the contact area of one lump and substituting Eq. (7) into the energy balance give now $Np = \eta'\omega WNs$. We get $W \propto 1$ that is the vibrational energy density is independent of the number of sources. The number of lumps is responsible of an increase of supplied power, but, in the mean time, it is also responsible of an increase of dissipation exactly in same proportion. This well explains the constant regime of sound observed on drum. When combining both dissipation mechanisms in material and contact, all intermediate regimes exist from constant law of sound to the linear law of 10 dB per decade [7].

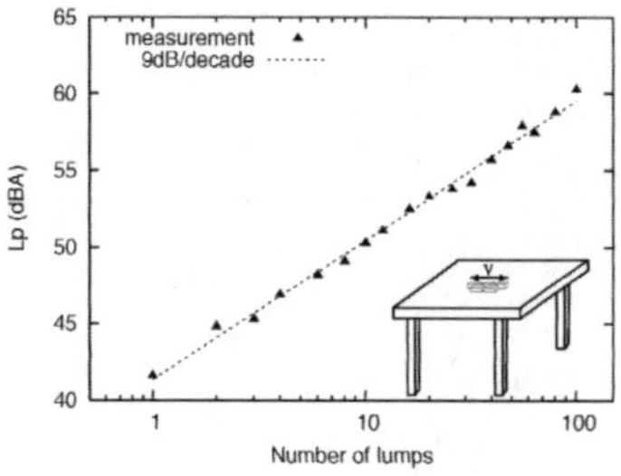

Figure 8. Friction noise generated by sugar lumps sliding on a wood table. The Sound Pressure Level (SPL) increase by 9 dB per decade (linear regime).

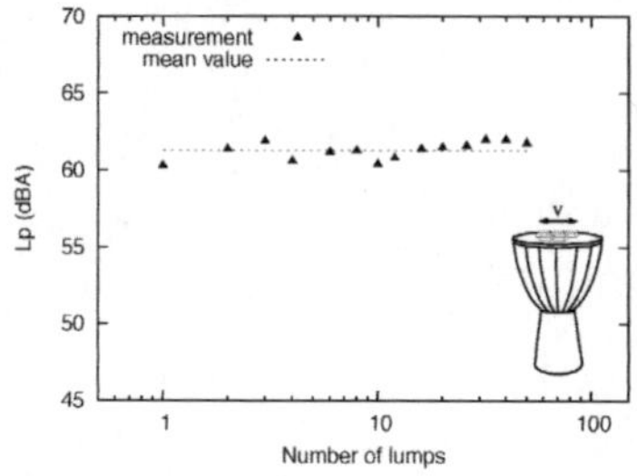

Figure 9. Same experiment on a drum membrane. The Sound Pressure Level (SPL) is almost independent on the number of lumps (constant regime).

6. Conclusion

Roughness noise is a very rich phenomenon. The mechanical problem of contact and vibration is highly multi-scale (scale of asperities much lower than scale of vibration). This is also a problem of statistical mechanics since the surfaces in presence are random by nature. Numerical simulation of contact mechanics is intensively time-consuming but highly optimized methods are possible. They lead to interesting results such as impact rate, impact duration, and strength of impacts.

The duality of contact, source of vibration but also dissipation of vibration, is certainly the most intriguing characteristic of roughness noise. The complexity of local events in contact does not allow a deterministic approach of the problem. But the fact that microscopic events responsible of noise emission are numerous, short and random is a favorable condition to apply methods and concepts of statistical mechanics.

Acknowledgments

This work was performed within the framework of the Labex Celya of Université de Lyon, operated by the French National Research Agency (ANR-10-LABX-0060/ANR-11-IDEX-0007).

References

[1] Akay A 2002 Acoustics of friction, *J. Acoust. Soc. Am.* **111** 1525-1548
[2] Stoimenov B L, Maruyama S, Adachi K and Kato K 2007 The roughness effect on the frequency of frictional sound, *Tribol. Int.* **40** 659?664
[3] Ben Abdelounis H, Le Bot A, Perret-Liaudet J, Zahouani H 2010 An experimental study on roughness noise of dry flat surfaces, *Wear* **268** 335-345
[4] Yokoi M and Nakai M 1982 A fundamental study on frictional noise (5th report?The influence of random surface roughness on frictional noise). Bull JSME **25**(203) 827?33.
[5] Rabinowicz E, 1951The Nature of the Static and Kinetic Coefficients of Friction, *J. App. Phys.* **22** 1373-1379
[6] Hung V H, Perret-Liaudet J, Scheibert J, Le Bot A 2013 Direct numerical simulation of the dynamics of sliding rough surfaces, *Comp. Mech.* **52** 1169-1183
[7] Le Bot A, Bou Chakra E, Michon G 2011 Dissipation of vibration in rough contact, *Tribol. Lett.* **41** 47-53
[8] Le Bot A 2015 *Foundation of statistical energy analysis in vibroacoustics*, Oxford University Press, Oxford
[9] Le Bot A, Bou Chakra E 2010 Measurement of friction noise versus contact area of rough surfaces weakly loaded, *Tribol. Lett.* **37** 273-281